CASE STUDIES IN MECHANICAL ENGINEERING

CASE STUDIES IN MECHANICAL ENGINEERING

DECISION MAKING, THERMODYNAMICS, FLUID MECHANICS AND HEAT TRANSFER

Stuart Sabol

WILEY

This edition first published 2016
© 2016 John Wiley & Sons, Ltd.

Registered Office

John Wiley & Sons, Ltd, The Atrium, Southern Gate, Chichester, West Sussex, PO19 8SQ, United Kingdom

For details of our global editorial offices, for customer services and for information about how to apply for permission to reuse the copyright material in this book please see our website at www.wiley.com.

Library of Congress Cataloging-in-Publication data applied for.

ISBN: 9781119119746

A catalogue record for this book is available from the British Library.

Cover image: GettyImages-157641784 – 77studio

Set in 10/12pt Times by SPi Global, Pondicherry, India
Printed and bound in Singapore by Markono Print Media Pte Ltd

1   2016

This book is dedicated to my wife. She is my companion, my support, greatest believer, and my best friend.

Contents

Foreword

Professors teaching engineering and corporate managers teaching entry level engineers will find this book an invaluable resource not found in any university curriculum. The author, Mr. Stuart Sabol, has drawn from his many years in engineering in industry and has enthusiastically written this book in an attempt to complement the engineering knowledge gained from a university curriculum with real complex system engineering problems that were actually encountered in the real world and that impacted both his career and the bottom line of the companies involved.

Stuart Sabol is an engineering expert who was not only intimately involved in but was pivotal to the solutions of some of the most critical and complex problems in large-scale system engineering. In most university engineering courses students are given the problems to solve with only the data required to solve them. This unrealistically hints at the correct solution. In the real world, however, an abundance of information is available or can be determined and thus good engineering judgement is required to determine what information is crucial. The author presents, through his many industry case studies, an abundance of information for each in terms of data, background, photos, and drawings from which a student may draw to determine the best course of action. Mr. Sabol has organized the case studies with a number of special exercises for students or for student teams to perform. The actual resolution for each practical case study is also given for discussion.

I think that the author can be confident that there will be many grateful professors, students and engineering managers who will have gained a broader necessary perspective of real world engineering and the associated multidiscipline approach required to solve the large scale problems frequently encountered in industry.

Dr. William C. Schneider*

Endowed Professor of Engineering
Texas A&M University
and (retired) Senior Engineer, NASA Johnson Space Center

---

*Dr. William C. Schneider is presently an Endowed Professor of Engineering, teaching courses in engineering mechanics and engineering design to graduate and undergraduate students at Texas A& M University. He has a wealth of practical experience gained from 38 years performing critical analysis and design at the NASA Johnson Space Center. Some of his spacecraft components are still on the moon; he has 14 US patents and numerous NASA medals for his achievements. He had sign-off authority for the US space-shuttle flights.

Preface

Being an engineer, husband and father rank highly among my endeavors. I have had great mentors throughout my professional and personal life; and have tried to be a good mentor to those I worked with, and to those close to me. This book is perhaps a completion of that attempt to mentor others to become better engineers.

When I ask someone, "How do you solve a problem?" they look at me and ask anything ranging from "Don't you know?" to "What problem?" I don't know the answer either. What I do know is that the more problems I solve the better I am at solving problems. Thus, experience is a valuable teacher. The trouble is, experience takes time.

Reducing the time to gain experience in real-world problem solving is therefore a goal of this work. I have taken from my career the most memorable projects. They are memorable because they were difficult. Memorable because I learned something from each one. Although they may seem difficult, there are paths through the data and seemingly unconnected points that reside in our engineering education. My hope is that these scenarios open doors to problem solving and life beyond the university that will pay dividends in the reader's career as a mechanical engineer.

The cases in this book are experiences, altered to avoid identification with any owner. Names are excluded. Locations are not mentioned and in many cases transposed across oceans to disguise the original project. It is not my intent to identify anyone but to present a situation that provides a learning opportunity. I anticipate that each chapter, including the problems and outside readings, can be completed in one week as part of a supplement to course studies.

There are people and institutions that have made this book possible, and I would like to acknowledge a few of them. To the contributors of artwork, Mitsubishi-Hitachi Power Systems Americas, EPRI, Doosan, ERCOT, ThermoFlow, Fram, Nooter/Eriksen, Atco, General Electric, Siemens, ASME, Crane, DeWalt, Dresser, Alstom, Triad Instruments, and owners that permitted the use of photographs, I am deeply grateful. Being able to show size, scale, and details of equipment characteristics is a valuable contribution. Thank you.

My engineering and professional mentors are too numerous to recall; however, a few deserve a mention. Charles, it was great to work with you and create a first of its kind. Keith, you directed a mentor/mentee relationship that changed the company, and protected it in a unique project that resulted in a considerable new opportunity. Who says Fortran is dead? Reid, "you can have only one first priority," "everyone has a contribution to make," and "there are only two decisions you make in your life" are valuable life lessons that will stay with me. Mike taught me how to appreciate everyone's opinion, to seek them out, and incorporate everyone's knowledge. George helped me understand that there is no greater joy than to enjoy what you do. Jo showed me how to progress the work and how to motivate people.

A special thanks to Steve Turns at Penn State. Your feedback was a breakthrough for this book. Also a special thanks to my publisher. Paul, thanks for believing in the book as much as I did, and in my ability to create it.

Stuart Sabol PE, PMP

Introduction

This volume of *Case Studies in Mechanical Engineering* strives to bring real-life experiences to students, recent graduates, and those seeking to continue their education either formally or on their own. These particular cases depart from traditional engineering case studies in that they are not evaluations of failures, and do not try to explore the field of engineering ethics. Instead, the author has drawn from his years of engineering to present those cases that affected his career and brought about new understandings in the field and practice of mechanical engineering. All deal with engineering's impact on a company's earnings and profit.

Each case is a study based on actual problems solved by engineers in industry. The names of the facilities and participants in the cases are absent, and the facts have been altered, but the lessons remain intact. Some of the case studies have been assembled from different projects or events that took place over several years. Most have been shortened or simplified to present a set of cases, each of which can be completed in a reasonable amount of time. The case studies thus provide a glimpse of how real-world engineering differs from traditional textbook problems and how engineering can impact management and the corporate bottom line.

Cases 1 through 3 are introductory. The first case study provides details of steam turbines, their design, and their operating characteristics. It provides a lesson in thermodynamic analysis, and its relevance to actual hardware. The second case study links commercial and engineering disciplines with the added dimension of time pressure and decision making. The third introduces manufacturer corrections from test to standard conditions for gas turbines combined with normal wear and tear, paradigm shifts, capital improvements and management decision-making processes.

Cases 4 and 5 explore aspects of detailed design. Case 4 studies the details of ASME flow elements in liquid and two-phase applications, thermodynamics, uncertainty evaluations, and computer programming. Case 5 dives deeper into applications of two-phase flashing flow with the problem of setting equipment elevations, pump characteristics, and detailed hydraulic calculations.

Case 6 develops a tool to analyze system availability and reliability.

Cases 7 and 8 deal with environmental subjects and an engineer's role in society and higher level decision making. Case 7 requires balancing of combustion calculations, and decision making. Case 8 explores fundamental market behavior and how a company's decisions can be impacted by taxes and governmental intervention.

Cases 9 and 10 deal with the application of engineering fundamentals combined with more abstract concepts. Case 9 combines knowledge of heat-transfer characteristics and detailed fluid system design, with quality-assurance requirements for engineers, and owners, and corporate responsibility. Case 10 develops a maintenance strategy for large equipment and complex systems, expanding the current state of the art in maintenance planning.

Case 11 explores the roles and responsibilities of an engineer responsible for a design team. The case illustrates developments, leadership and management tools that can be applied to a generic project engineering assignment.

Case 12 is a short study of engineering in daily life: how advancements are possible even in trades that have evolved over millennia, highlighting the necessity of using creativity and improving accuracy and quality.

In these studies, career decisions, standard practices, and engineering improvements are combined with decision-making and presentation skills to advance the traditional textbook approach to engineering beyond the classroom.

Each case study contains several exercises that can be used as in-class or homework activities. The cases may be approached as team activities. Solutions to the exercises, and detailed discussion, are included in each chapter.

Case 1

Steam Turbine Performance Degradation

A private investor-owned power company owns 15 GW of capacity including conventional fossil-fired generation and natural-gas fired combined cycle gas turbine power plants spread throughout the United States. The company competes in several unregulated power markets and takes seriously its ability to provide safe, reliable, low-cost power compared to its competitors while meeting all environmental permit requirements. Quarterly senior management reviews include reports on worker and contractor safety performance, the reliability and efficiency of the facilities, as well as any exceedances of environmental permits. The company spent time and resources establishing guidelines and procedures for regular performance monitoring at its generating facilities, including results analysis. These guidelines are routinely reinforced at every level of the organization with training for new recruits and refresher courses for midlevel management.

The performance-monitoring procedures and guidelines include techniques to analyze the test data based on industry guidelines, particularly ASME PTC Committee (2010) and technical papers from noted industry experts such as Cotton and Schofield (1970). For the company's steam turbines, the condition of the various stages is related to changes in stage pressures at standard conditions knowing how the throttle flow to the machine has changed. The methods are based on the fact that, for a large multistage condensing turbine, all stages, except the first and last, operate with a constant pressure ratio $(p_2/p_1.)$ This allows the general flow equation for flow through a converging-diverging nozzle for stages beyond the first stage to be simplified to equation (1.1)

$$\dot{m} = \Phi \cdot \sqrt{\frac{P}{\upsilon}} \tag{1.1}$$

Case Studies in Mechanical Engineering: Decision Making, Thermodynamics, Fluid Mechanics and Heat Transfer, First Edition. Stuart Sabol.
© 2016 John Wiley & Sons, Ltd. Published 2016 by John Wiley & Sons, Ltd.
Companion website: www.wiley.com/go/sabol/mechanical

where

> $\dot{m}$, P and v are the flow rate, absolute pressure and specific volume to the following stage;
> Φ is a constant flow function (area).

The flow function Φ includes unit conversions, constants of proportionality, the area of flow, and the coefficient of discharge for the nozzle and blade path. Except for unit conversions it has units of area.

A production engineer at one of the company's coal-fired power plants with three 600 MW subcritical single reheat units has been monitoring the units' performance according to company procedures. In just over 7 months since the last major overhaul one unit has lost 3.4% of its output, and the cycle heat rate has increased 0.6%. Using the guidelines, most of degradation in performance can be explained by changes in the flow-passing capability of the steam turbine and losses in the high-pressure (HP) turbine efficiency.

However, there are changes to characteristics that are not discussed in the corporate standards or the technical papers available in the office. In particular, the intermediate pressure (IP) turbine's extraction temperature has risen noticeably from the expected value. Efforts to explain the symptoms as instrumentation issues have failed. Rather than dismiss or ignore the findings, you, the engineer, are determined to find the cause, its economic value, and to recommend a course of action to address the issue.

1.1   Steam Turbine Types

The variety and application of steam turbines is enormous. It includes the utility tandem compound unit pictured in Figure 1.1, mechanical drives for onshore or marine applications, combined-cycle and single Rankine-cycle units, super critical, single or double reheat units, and nuclear power-plant applications. One way to categorize the various models is by size. Very basically, smaller installations typically serve as variable speed mechanical drives for pumps and compressors. These may be as large as 50 to 75 MW and have inlet conditions up to 750 psi (5.2 MPa) and 700 °F (644 K). Many are located within chemical processing plants or refineries and exhaust into a lower pressure steam header that provides steam for heating, or to drive smaller steam turbines that may exhaust into a surface condenser. The larger varieties will be multistage units with an axial flow exhaust.

Up to about 150 MW, steam turbines typically have an axial flow exhaust with throttle conditions as high as 1500 psi (10 MPa) and 900 °F to 1000 °F (755 K to 810 K). Figure 1.2 shows a drawing of a Siemens axial flow machine. Such turbines may be used in a chemical process plant and have a controlled extraction for process heat or other uses. This size is also common in combined cycle power plants with uncontrolled expansion to the condenser. Occasionally, an axial flow machine will have single reheat as part of the cycle. If it is a condensing cycle, the condenser can be placed on the same elevation as the turbine. Combined cycle units utilize waste heat from a gas turbine to generate steam; thus, steam-turbine extractions for regenerative heating are not employed in a combined cycle. A single Rankine cycle would employ uncontrolled extractions for feedwater heating.

Above approximately 150 MW, the last stage blade (L-0) becomes too long to manufacture and operate reliably. The low pressure (LP) turbine becomes a dual flow design with steam entering the center section and steam traveling in opposing directions to exhaust downward

Figure 1.1   Alstom steam turbine. *Source*: Reproduced by permission of Alstom.

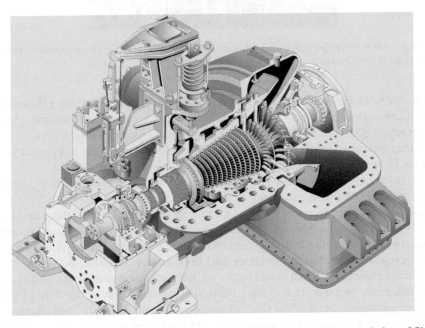

Figure 1.2   Typical axial flow exhaust steam turbine. *Source*: Reproduced by permission of Siemens Energy.

Figure 1.3   700 MW ST Hekinan Unit 3, Chubu Electric Power Co. *Source*: Reproduced by permission of Mitsubishi Hitachi Power Systems America, Inc.

into the condenser. For these machines, the steam turbine must be raised above the condenser, which increases construction costs to include foundations for an elevated turbine. Figure 1.3 is a photograph of the 700 MW ST Hekinan Unit 3, Chubu Electric Power Co. steam turbine. The tandem compound machine has dual flow HP and IP sections in the foreground with two dual-flow LP sections in the background.

Machines as large as 650 to 750 MW usually operate with subcritical steam pressures with a single reheat. Throttle conditions may be as high as 2800 psi (19 MPa) and 1050 °F (840 K) with the reheat temperature matching the throttle temperature. Units in this size range are generally uncontrolled expansion, condensing units used for power generation either in combined cycles or single Rankine cycle units with regenerative heating. The larger single Rankine-cycle units may have two or three dual-flow, down-exhaust LP sections. Combined-cycle steam turbines are limited in size by the gas turbine portion of the combined cycle. As a rule of thumb, the steam portion of the combined cycle plant is about one-third of the plant total electrical output. Most of the single Rankine-cycle units are fossil fired although some may be in nuclear facilities. Combined cycle and fossil-fired units operate at synchronous speed with a two-pole generator. Nuclear units typically have four-pole generators and operate at half synchronous speed.

Above about 650 MW, fossil-fired units begin using supercritical pressures and may include double reheat Rankine cycles with regenerative feedwater heating. Throttle conditions may be above 4000 psi (28 MPa) and 1150 °F (895 K). Reheat temperatures usually match the throttle conditions but cost optimizations may result in the reheat temperatures somewhat above the throttle. Nuclear steam cycle conditions generally do not change much with size. The largest steam turbine at the time of this writing was in the neighborhood of 1800 MW.

1.1.1 Steam Turbine Components

The active components of steam turbines are the rotating and stationary blades. Rotating blades are sometimes referred to as *buckets*, from their shape. Steam to the machine is controlled by multiple throttle valves. In large modern machines there are four hydraulically controlled valves that can close very quickly in the event of an upset. From the control valves, the steam is directed to the first control stage through sets of nozzles. Each set of nozzles accepts steam from one of the inlet throttle valves. The first control or governing stage has impulse or Curtis blading. Beyond the governing stage, the blades take on an increasing amount of reaction, as the pressure diminishes and the pressure ratio increases across each rotating stage.

The rotating blades are mounted on wheels or disks that are fixed to the shaft, or the shaft is machined with integral wheels to accept the blades – see Figure 1.4. The wheels provide increased torque on the shaft. The blades are secured in the wheel by a dovetail or fir tree shaped slot. Each blade is weighted and moment balanced, then ordered so that the assembled rotor is nearly balanced. During assembly of the rotor, the blades are slid into the dovetail slots until the ring is full. The blades are locked in place and the locking mechanism is frequently peened to ensure the security of the blades during operation.

A large steam-turbine generator in a reheat cycle will have a high pressure (HP) rotor, one or more intermediate pressure (IP) rotors, and one or several low pressure (LP) rotors. These may be mounted on a single shaft (tandem compound) or in a dual shaft arrangement with two

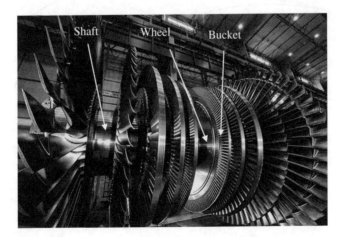

Figure 1.4 An LP section of a large nuclear steam turbine. *Source*: Reproduced by permission of Alstom.

generators (cross compound). Intercept valves are located in the hot reheat line immediately upstream of the IP turbine section. These valves do not control flow but are configured to close quickly in the event of an upset condition preventing the energy stored in the steam lines to and from the boiler reheater from overspeeding the rotor after the generator is disconnected from the electric grid. The pipeline to the boiler from the HP turbine exhaust is referred to as the "cold" reheat line. From the boiler to the IP turbine is the "hot" reheat line.

Once assembled, each rotor is balanced first at slow speed then at high speed. If a field repair requires replacement of worn or damaged blades, the rotor may be removed from the casing and field balanced prior to completing the repair. A trim balance may also be required once the machine is assembled and run at full speed for the first time.

Between rotating blades are stationary vanes, sometimes referred to as nozzles or dia-phragms. The blades of the diaphragms are shop assembled in halves between inner and outer blade rings. One half will be fitted in the lower casing and the other half in the top half of the turbine casing. The stationary vanes turn and focus the steam from the exhaust of the upstream rotating blade and direct it to the downstream blade. Each expansion stage is a combination of one set of nozzles and one set of blades.

During assembly of the rotors and casings, sealing strips are mounted to the rotor between the wheels, and in the casing between diaphragms. Mating seal strips are placed on the blade tips, or are made as an integral part of the blade, to match with the casing seals. The inner rings of the diaphragms have seals that match with the rotors. Figure 1.5 shows blade tip seals from US Patent 6926495. The sharp edges are an inefficient flow path reducing leakage between the rotating and stationary components. The sealing components are adjusted during assembly so that there is proper clearance throughout the circumference to prevent the rotating seals from striking the stationary components in the cold and operating conditions.

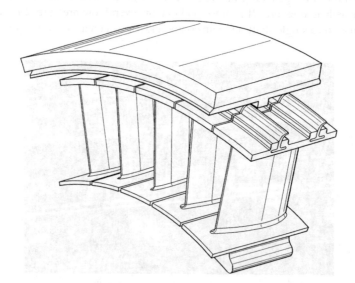

Figure 1.5   Blade tip-seals (US Patent 6926495 Ihor S. Diakunchak). *Source*: Ihor S. Diakunchak, Siemens Westinghouse Power Corporation.

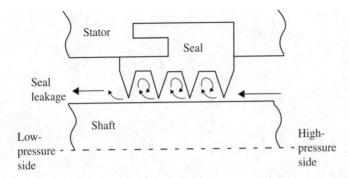

Figure 1.6 Shaft seal leakage. *Source*: Reproduced by permission of Dresser-Rand Company.

The ends of each shaft have a series of labyrinth seals to reduce the amount of high-pressure steam leaking out and air leaking in at the ends of the LP rotors – see Figure 1.6. Very often the HP and IP turbine sections are built on a single shaft and separated by a labyrinth seal between the HP exhaust and IP inlet. High-pressure steam-seal leakage is collected and used as sealing steam in lower pressure sections. It is also injected into the LP shaft seals to prevent air in-leakage.

Once the lower half casing is loaded with diaphragms, seals and bearings; the rotors are lowered into place onto the bearings. The rotors are positioned and aligned with the casing, and mated to the other tandem shafts and the generator rotor. Once all clearances are checked, the upper half bearing halves installed, and the upper half diaphragms and seals in place, the upper casing is lowered onto the bottom casing, and the two casing halves are bolted together. Each bolt is torqued, then heated and torqued again so that it provides the proper tension at operating temperature.

Figure 1.7 shows a tandem compound machine with a dual flow LP casing and the rotor in place inside the lower casing. Stationary vanes in the lower casing can be seen upstream of the latter stages. There are multiple steam extraction points along the flow path.

Steam and auxiliary piping, and instrumentation are assembled prior to insulating the casings and piping segments around the turbine. If there are turbine housings, these are assembled for normal operation.

1.1.2 Startup and Operation

Once assembled, the turning gear motor maintains a slow shaft rotation of about 4 to 5 rpm to prevent bowing. Bowing of the rotor under its own weight during assembly or by differential cooling when the machine is temporarily off line can cause high vibrations during startup. Slow rotation of the shaft prevents differential cooling from bowing the rotor and reduces bowing due to maintenance enough to allow a safe roll up to operating speed. Prior to startup, any water in the casing from steam condensation must be drained to prevent impact damage on the rotating blades.

The high-pressure casing of a large steam turbine can be 10 to 12 inches thick; thus, the rotor will reach operating temperature much faster than the shell. To avoid differential thermal expansion between the rotor and shell, steam turbines must be warmed slowly to prevent

Figure 1.7 Tandem Compound steam turbine with extractions. *Source*: Reproduced by permission of Doosan Skoda Power.

contact between the rotating and stationary components. The manufacturer specifies heating rates and allowable ranges for differential expansion, which are continuously measured and maintained within allowable limits during startup and operation.

In recent years, manufacturers have devoted considerable efforts to decrease the warm-up time for steam turbines in combined cycles. Fast starting units on a single shaft with a gas turbine driving a single generator are less expensive to build and can provide backup generation for renewable power sources. See the web sites below for more information:

https://powergen.gepower.com/services/upgrade-and-life-extension/heavy-duty-gas-turbine-
  upgrades-f-class/power-flexefficiency.html (accessed January 25, 2016).
http://www.energy.siemens.com/hq/pool/hq/power-generation/power-plants/gas-fired-
  power-plants/combined-cycle-powerplants/Fast_cycling_and_rapid_start-up_US.pdf
  (accessed January 13, 2016) (Baling, 2010).

After steam is raised in the boiler, sealing steam provided to the LP shaft seals allows evacuation of air from the condenser. Pulling vacuum permits introduction of steam through the throttle valves, disengaging the turning gear, and accelerating the shaft up to rated speed. During the rollup, condensing steam from contact with cool turbine parts, and that which condenses through the expansion path, must be continually drained from the casing to avoid damage that can be caused by sudden vaporization or impact with rotating components.

There will also be several critical speeds during rollup when the shaft resonates, causing spikes in vibration. These speeds are anticipated, and the operator will accelerate the shaft through the critical speeds rather quickly. If vibrations exceed allowable limits, the startup is aborted and the shaft allowed to coast down before being placed on turning gear. If this happens, the unit may remain on turning gear for a manufacturer-recommended period prior to a second attempt to start. If the situation warrants, a trim balance weight may be calculated and placed on the shaft.

As the shaft rotation approaches full speed, the operator will begin preparations to engage the driven equipment or place the process in normal operation. In the case of a large steam turbine generator, the operator will prepare to close the generator breaker. Once the breaker is closed, load on the generator is rapidly increased by increasing steam flow, pressure and temperature through the throttle valves. At specified temperatures, the casing drains are closed, and the unit begins normal operations. Normal operation may continue without interruption for years!

During normal operations, the staff must diligently maintain feedwater and steam purity. Several elements or compounds can vaporize at subcritical boiler pressures, most notably silica. Silica present in sufficient concentration will vaporize in the HP boiler drum and condense in the HP turbine as the steam temperature drops through expansion. Copper can vaporize as well and carry through to the HP turbine. Spray water used to control main steam and reheat temperatures will also inject whatever is in the water directly into the downstream turbine sections.

In supercritical units, there is no boiler drum; thus feedwater quality is more critical. Anything in the water will be carried directly into the HP turbine section.

Reading

Read Cofer et al. (1996). To test your understanding of this article, answer the following questions. An answer key is provided at the end of this case study.

1. What components of the blade path design account for up to 90% of the efficiency losses?
   (a) Aerodynamic design and tip leakage.
   (b) Rotational losses.
   (c) Shaft packing leakage.
   (d) Inlet nozzle losses.
2. What is a logical first step in the design of an improved blade path?
   (a) Laboratory wind-tunnel tests.
   (b) Computational fluid dynamic calculations.
   (c) Full-scale wind-tunnel tests.
   (d) Customer field testing in commercial applications.
3. Why is the last stage bucket design the most important contributor to performance?
   (a) From the First Law: work is equal to $\int P \, dv$.
   (b) The last stage bucket operates in the wet region.
   (c) The last stage bucket produces 10% to 15% of the steam turbine output.
   (d) Answers a and c above.

1.1.3   Performance Monitoring and Analysis

Due to the large pressure drop from the throttle to the condenser (1.7 to 12 kPa) the general equation for flow through a converging-diverging nozzle can be simplified considerably. The general flow equation for subsonic nozzle flow requires knowledge of the upstream and downstream pressures. For a large steam turbine with uncontrolled expansion, except for the first and last stages, the pressure ratio across the stages is constant throughout the load range. Therefore, the downstream pressure is proportional to the upstream pressure, and the general flow equation can be reduced to equation (1.1).

The first stage of the turbine is affected considerably by changes in flow, and the last stage is affected by flow and condenser pressure.

With the simplified flow equation, much can be determined about the condition of a turbine by monitoring feedwater flow, and pressures and temperatures into and out of the various turbine sections (HP, IP and LP inlet). The HP turbine first-stage pressure is measured downstream of the governing row or rows.

Not long ago, scheduled and carefully controlled performance tests with precision instruments were the best way of monitoring steam-turbine performance. Thermocouples calibrated against standards in precision furnaces measured with precision potentiometers and dead weight gages for pressure measurement were the instruments of choice. With the advances in digital computing and in temperature and pressure transmitters, online monitoring has become the norm. Regular calibrations of the transmitters and use of transmitters that self-correct for ambient temperature can yield sufficiently accurate results. Automated collection of instrument readings and computer storage systems retain data for long periods permitting research into performance trends that can help determine the cause or causes of performance deterioration. Furthermore, the data are available at the site, in a corporate office or easily made available to consultants. Acceptance tests can still require special test instrumentation, but even these are more often conducted with mostly plant instrumentation calibrated just prior to the test.

1.1.4   Analyzing Performance Data – Corrected Pressures

With the data, there are two basic methods for analyzing performance trends. The first relies on corrected stage pressures with knowledge of changes in throttle flow. This method requires the engineer to correct the measured output and measured pressures to standard conditions. Manufacturer curves provide the means to correct the unit output to reference or design conditions. The first-stage pressure correction is shown in equation (1.2) from Cotton and Schofield (1970):

$$P_C = P_O \cdot \left(\frac{P_{dt}}{P_{Ot}} \right) \tag{1.2}$$

Extraction pressures, if present in the HP expansion path, may be corrected with the same correction without loss of accuracy.

Pressures beginning with the Hot Reheat Intercept Valve and downstream are corrected with equation (1.3), again from Cotton and Schofield (1970):

$$P_C = P_O \sqrt{ \frac{P_{dt}}{P_{Ot}} \cdot \frac{\upsilon_{Ot}}{\upsilon_{dt}} } \sqrt{ \frac{\upsilon_{dr}}{\upsilon_{tr}} } \tag{1.3}$$

where:

P_C = corrected pressure (absolute);
P_O = observed pressure (absolute);
P_{Ot} = observed throttle pressure (absolute);
P_{dt} = reference throttle pressure (absolute);
v_{dt} = reference throttle specific volume;
v_{Ot} = observed throttle specific volume;
v_{tr} = test specific volume at the intercept valves; and
v_{dr} = reference specific volume at intercept valves.

Use of corrected pressures to analyze the condition of the steam turbine is best explained by example. In a particular case, the following changes in observed parameters occurred over a 22 month period:

Example 1.1

Throttle flow −17.2%
Output −16.5%
First-stage pressure +21.2%
High-pressure efficiency −12.2% (Cotton and Schofield, 1970)

The lower throttle flow together with the high first-stage pressure indicates loss of ability to pass steam in the second or latter stages of the HP turbine. As the degradation in performance occurred over a period of 22 months, the cause was suspected to be deposits rather than mechanical damage or components that became lodged on blades or vanes. The inspection revealed heavy deposits throughout the HP turbine.

When throttle flow is not available, analysis is a bit more complicated. However, logical analysis of the data generally yields accurate interpretations. For example, again from Cotton and Schofield (1970), the following data from a turbine efficiency test showed significant changes from a previous test eight months earlier.

Example 1.2

Output −2% (approximate)
First-stage pressure +0.3%
Hot-reheat pressure +6.3%
Low-pressure inlet +2.2%
HP turbine efficiency −1.3%
IP turbine efficiency −6.9%

The large increase in the hot-reheat pressure without a corresponding increase at the first stage indicated that there was a flow restriction at the intercept valves or downstream in the IP turbine. Some plugging might be indicated in the LP turbine as well. The intercept valves were eliminated by a quick measurement of the pressure loss across the valve; therefore, pluggage in the IP turbine was suspected. Deposits in the IP turbine but not in the HP turbine would be possible, but not likely. Since the first-stage pressure changed very little, mechanical damage in the IP turbine was suspected.

Upon inspection, distortion of the IP casing was found, which caused severe rubbing at the ninth stage, which liberated blade covers, and resulted in failure of ninth stage blades. The failed blades further damaged the ninth-stage vanes, which were removed from the diaphragm, and caught between the rotating blades and the vanes of the ninth stage. The distortion was determined to be caused by water induction into the operating unit that suddenly cooled a portion of the shell.

Example 1.3: Check Knowledge

What do the following symptoms indicate from the corrected conditions shown below?

Power output −7%
First-stage pressure +2%
Hot-reheat pressure −3%
LP inlet pressure −9%
HP turbine efficiency −7%
IP turbine efficiency −1.5% (Cotton and Schofield, 1970)

Analysis

The rise in first-stage pressure indicates either an increase in flow or decreased flow-passing ability in the remainder of the HP turbine. If flow had increased, due to increased first-stage area, the increased flow would have been expressed throughout the remainder of the flow path. The IP and LP inlet pressures decreased, so an increase in flow is eliminated.

Therefore, the most probable cause is some type of blockage in the second and subsequent stages of the HP turbine. Since the LP inlet pressure fell substantially more than the IP inlet, there is likely additional blockage in the IP turbine, which is supported by the drop in IP turbine efficiency.

The information in Example 3 does not specify the time over which the changes took place. Nevertheless, there should be some common connection between the simultaneous changes in the HP and IP turbine performance. This could be due to deposits from the steam or another mechanism, such as scale that abrades the blades and vanes of the HP and IP turbines. Deposits generally lead to increased stage pressures. Only the first stage pressure increased here, so a mechanism of mechanical abrasion or mechanical damage to the blades should be suspected. Inspection of the machine showed that solid particle erosion had removed considerable material from the blades in the HP and IP turbines. Figure 1.8 shows an example of vanes heavily eroded by solid particle abrasion.

1.1.5 Analyzing Performance Data – Flow Function

The second method of analyzing the turbine performance is to go directly to a calculation of Φ, a method that does not require corrections to standard conditions. Instead thermodynamic states are determined at the inlets to main turbine sections, and an energy and mass balance determines flows throughout the cycle. Changes in the value of Φ imply physical changes inside the turbine such as flow area, surface condition, or shape of the blades in the following

Figure 1.8   Solid particle erosion. *Source*: Wilcox *et al.* (2010).

stages. If deposits clog the area, Φ will decrease. If the blade surface becomes rough, if a turbine component or foreign object blocks the flow path, or if mechanical damage changes the shape of the blades or vanes, Φ will decrease. An increase of the downstream area due to erosion or mechanical damage will show an increase in the value of Φ.

Example 1.4

Example 1.1 above, suppose the steam turbine has the following design conditions:

Turbine inlet condition:
    pressure (psia / MPaa): 3514.7 / 24.233
    throttle temperature (°F /K): 1050 / 838.7
Governing stage exit condition:
    first-stage pressure (psia / MPaa): 2612.1 / 18.01
    first-stage enthalpy (Btu/lb / kJ/kg): 1431 / 3329

At these conditions, the apparent governing stage efficiency from the throttle valves is 72%. The specific volume at the first stage is 0.281 ft³/lb (0.0175 m³/kg). From the data in the example, the first-stage pressure increases 21.2% to 3151 psig (21.725 MPa). Assuming the first-stage efficiency falls at the same rate as the overall HP turbine efficiency yields enthalpy and specific volume values of 1451 Btu/lb (3375 kJ/kg), and 0.2397 ft³/lb (0.0150 m³/kg), respectively, at the corrected test conditions. This assumption is not necessarily accurate but is a fair approximation.

A change in the flow function can be calculated from equation (1.4):

$$\frac{\Phi_{2O}}{\Phi_{2d}} = \sqrt{\frac{P_{1d}}{\upsilon_{1d}}} \left(\frac{\dot{m}_{1t}}{\dot{m}_{1d}} \sqrt{\frac{\upsilon_{1O}}{P_{1t}}} \right) \tag{1.4}$$

where:

Φ_{2d} = design second-stage flow function;
Φ_{2O} = observed second-stage flow function;
$\dot{m}_{1d}$ = design flow to the first stage;*
$\dot{m}_{1t}$ = observed flow to the first stage;*
P_{1d} = design first-stage pressure (abs);
P_{1O} = observed first-stage pressure (abs);
v_{1d} = design first-stage specific volume; and
v_{1O} = observed first-stage specific volume.

For this example, the ratio of test to design flow function to the stage following the governing stage is 0.694 indicating a diminished flow passing ability of second or subsequent stages of the HP turbine. As with the original analysis, the steady decline over a 22-month period would indicate deposits on the blades of the second or subsequent stages of the HP turbine.

The flow function at the HP turbine inlet (at the throttle valves), in this example, changes in the same proportion as the throttle flow (−17.2%). A calculation of Φ at the first stage exhaust shows that there is a progression of the deposit buildup through the steam path. In other words, the deposits should be expected to become greater and greater as the steam passes through the HP turbine.

Similar analysis is possible at the IP and LP turbines provided there is energy balance information available showing the flows to the following stages.

1.2   Refresher

1.2.1   Steam Turbine Efficiency

Steam turbine efficiency is calculated from the ratio of the actual change in enthalpy between the inlet and outlet to the isentropic change in enthalpy between the same two states. Turbines operating in the superheat region, dry steam, have the thermodynamic states determined by measured pressure and temperature. In the wet region, the thermodynamic states must be determined by heat balance or another means of determining the thermodynamic state. When calculating turbine section efficiency, the inlet thermodynamic state is usually downstream of the throttle or intercept valve, if there is one.

As an example, consider an HP turbine with throttle conditions of 31.02 MPa, and 866.5 K, exhaust conditions of 7.76 MPA and 650.8 K and a 4% pressure drop through the throttle values. The ideal and actual expansions are shown in the Mollier diagram of Figure 1.9. The section efficiency is 86% between an inlet enthalpy of 3416 kJ/kg and an exit of 3081 kJ/kg.

1.2.2   Example

Determine the exhaust temperature of an HP turbine section with throttle conditions of 16.68 MPa, 810.9 K, a 4% throttle valve pressure drop, an exhaust pressure of 4.97 MPa, and a section efficiency of 84%.

---

*Throttle valve stem leakage and seal leakage from the first-stage yields a flow to the second and subsequent stages of the HP turbine that is less than the throttle flow. For the examples in this case study, throttle flow is used as a reasonable approximation.

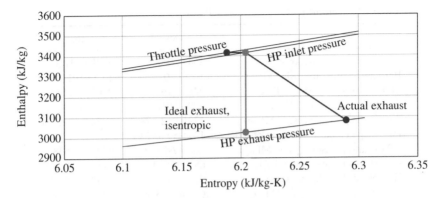

Figure 1.9 Example: HP expansion.

Solution:

Throttle enthalpy: 3398.2 kJ/kg
HP inlet pressure: 16.01 MPa
HP inlet entropy: 6.432 kJ/kg-K

Isentropic exhaust enthalpy: 3077.5 kJ/kg
Exhaust enthalpy: 3110.4 kJ/kg
Exhaust temperature: 638.7 K

1.3   Case Study Details

A nominal 600 MW fossil-fired subcritical, single reheat steam power plant has seven stages of regenerative feedwater heating. All extractions are uncontrolled, governed only by the heat transfer to the feedwater. The first steam extraction point (in the direction of steam flow) is taken from the HP turbine exhaust, the cold reheat line. The second is taken from an extraction port about midway through the IP turbine and the third at the IP turbine exhaust. The remaining four extractions are taken from extraction ports in the two identical dual-flow, downward exhaust LP turbines, which exhaust to the condenser at about 1.5 "Hga (5.08 kPa) at design conditions.

1.3.1   Performance Trend

In a little over 7 months, load has dropped 3.4%. Regular turbine performance tests on the unit with the throttle valves wide open (VWO) show the following changes after seven months:

Throttle flow: −6.31%
First-stage pressure: +3.04%
HP turbine efficiency: −1.25%
IP turbine efficiency: slightly less than expected

A performance test after 7 months showed the uncorrected pressures and temperatures given in Table 1.1.

Table 1.1 Turbine performance data – Part 1.

	Design	Test
Throttle flow (kpph / kg/s)	3641 / 458.8	3423 / 431.3
Barometric pressure (psia / kPaa)	14.7 / 101	14.7 / 101
Throttle pressure (psig / MPag)	2400 / 16.547	2405 / 16.582
Throttle temperature (°F / K)	1000 / 810.9	998 / 809.8
First-stage pressure (psig / MPag)	1795 / 12.378	1854.2 / 12.784
First-stage enthalpy (Btu/lb / kJ/kg)	1431.5 / 3329.7	
HP exhaust pressure (psig / MPag)	706.8 / 4.873	672.7 / 4.638
HP exhaust temperature (°F / K)	682.2 / 634.4	671.6 / 628.5
Hot reheat flow (kpph / kg/s)	3304 / 416.3	3142 / 335.8*
Hot reheat pressure (psig / MPag)	634.6 / 4.376	603.9 / 4.164
Hot reheat temperature (°F / K)	1000 / 810.9	1003 / 812.6

Assignment 1

From energy and mass balance calculations following the test:

1. Calculate the corrected first-stage and hot-reheat pressures and determine a likely scenario for the lost output.
2. Calculate the first-stage thermodynamic state at test conditions. Assume the HP turbine section efficiency from the first stage to the exhaust degrades proportional to the degradation of the overall HP turbine efficiency.
3. Use the design and measured conditions to calculate the items below for design and test conditions:
   (a) Turbine inlet flow function.
   (b) First-stage turbine flow function.
   (c) IP turbine inlet flow function.
4. What do the parameters from step 3 above indicate about the condition of the HP and IP turbines?

1.3.2 IP Turbine Enthalpy Drop

In addition to the parameters shown above, Table 1.2 provides data to determine the IP turbine enthalpy drop.

Assignment 2

1. Plot the IP turbine expansion line on a Mollier (h-s) diagram. Use a 2% pressure loss through the intercept valves and a 3% pressure loss from inside the turbine shell to the extraction pressure measurement.
2. Calculate the overall IP turbine efficiency.
3. Calculate the two IP turbine section efficiencies – from the Hot Reheat to the second extraction and from the second extraction to the IP Turbine Exhaust.
4. Are the two IP section efficiencies from 3. above reasonable?

Table 1.2 Turbine performance data – Part 2.

	Pressure (psig / MPag)	Temperature (°F / K)
Hot reheat	603.9 / 4.164	1003 / 812.6
Second extraction	369.8 / 2.550	902.4 / 756.7
IP-LP crossover	211.6 / 1.459	745.4 / 669.5
	Design (kpph / kg/s)	Test (kpph / kg/s)
Expected extraction flow rate	140 / 17.6	131 / 16.5

5. The IP turbine expansion line is usually drawn as a straight line on the Mollier diagram from the point downstream of the intercept valves to the crossover. Calculate the temperature difference between the expected and measured extraction temperatures.
6. What are the most likely components or operating parameters that lead to the high extraction temperature?

Read Pastrana *et al.* (2001).

1.4 Case Study Findings

The observed difference between the expected and measured extraction temperatures were quickly verified as accurate by comparing the measured extraction temperature at the turbine extraction port to the temperature measured downstream at the feedwater heater. The test thermocouple was checked against another test thermocouple, both measured with a laboratory potentiometer. Given the slight change in IP turbine efficiency, liberation of parts was discounted. Therefore, the high extraction temperature was most likely due to seal leakage between the casing and rotating blades.

The machine in question was occasionally washed to remove suspected deposits on the HP turbine blades. Washes were conducted under low load conditions with the main steam pressure reduced and the main steam temperature controlled to just above saturation conditions via spray water. As the steam expanded through the HP turbine, the latter sections would operate in the wet region and much of the deposit could be removed. Casing drains carried the condensate and deposits to waste.

As the steam condensed during the short wash period, the rotor could cool much faster than the shell under condensing heat transfer causing high-differential expansion / contraction between the shell and rotor. While operators watched the parameters very closely, differential expansion resulting in contact of the sealing surfaces was a likely cause for the high seal leakage after just a few months' operation. Other causes may have included: normal wear and tear, high vibration during a startup or wash, or improper installation during the last major overhaul. Normal wear and tear was eliminated given the short operating period during which the seal leakage increased. Therefore, the three most probable causes for the seal leakage were high differential expansion, high vibrations, and improper installation.

Measurements through the LP turbine expansion combined with offline computer-based heat and material balances of the full steam cycle showed that the LP turbine suffered similar differences between the expected and measured extraction temperatures. The temperature differences were not systematic; that is, they did not progress in any pattern through the machine and were often negligible. Therefore, installation errors may have been the initial cause for the high seal leakage but this could not be determined at the time.

As identified earlier, the turbine inlet flow function changed proportionally to the change in throttle flow. This indicated that the deposits were present on the first-stage nozzles and blades. Given that the steam to the HP turbine was slightly superheated for the washes, the first stage did not receive the full benefit of wet steam washing. Therefore, the turbine washes were less and less effective leaving the company little choice but to remove the unit from service for a complete overhaul.

Inspection revealed heavy brown deposits throughout the HP turbine. Analysis of the deposits showed a high concentration of copper, which was the material of choice, at the time the unit was constructed, for heat exchanger tubes within the condenser and low pressure feedwater heaters. Highly pure condensate slowly dissolved the copper tubes and transferred the ions to the boiler where they vaporized with the steam. While minute quantities were dissolved from each tube, the great number of tubes and the effect of boiling led to concentrations in the boiler drum high enough to vaporize copper, which deposited in the turbine. Most of the copper was removed from the system in the HP turbine with little effect on the IP or LP turbines.

When the machine was removed from service several years later for its major overhaul, the IP turbine seals were found to be worn, and there was evidence of a rub over a portion of the arc.

1.5   Decision Making and Actions

1.5.1   Value

Prior to the machine inspection, there were several unanswered questions and uncertainties:

- Was the damage to the seals caused by high vibration or differential expansion?
- Was there an installation error?
- What is the value of the seal leakage?
- How can it be prevented in the future?
- What should be done now and who should do it?

In general, these types of questions appear over and over during a career. Basically these can be summarized below:

- Do uncertainties or unknowns need to be eliminated or reduced in order to make a decision?
- Is a decision necessary now?
- How does the condition affect the company's bottom line?
- Is there value in making a change?
- How should the change be implemented?

The first step in the process of answering these types of questions is to identify the value of the loss or losses. Perfection is generally not required – a Pareto analysis (80 : 20) may often be enough. Once determined, the value will help establish how strongly to pursue corrective actions. It will also help prioritize your own work schedule.

Constant pressure lines have just a slight upward curvature on the Mollier diagram in the superheat region, so they can be very accurately approximated by straight lines over small distances. With this assumption, the expected turbine shell enthalpy at the extraction point can be located by the intersection of two straight lines – one the line drawn from the IP inlet downstream of the intercept valve to the IP turbine exhaust, and the other a line connecting the entropy / enthalpy points along a constant pressure line at the turbine extraction shell pressure. The hot reheat entropy and the turbine exhaust entropy are usually close enough to one another for an accurate calculation of the extraction shell entropy and enthalpy. The intersection of these two lines yields an enthalpy generally less than 0.1 Btu/lb (0.2 kJ/kg) from the true intersection of the constant pressure line and the turbine expansion line. Using the foregoing assumption, just a few steam table calls in the turbine monitoring program allow a fast and efficient method of monitoring the expected and measured temperatures continuously at the second extraction point.

Assignment 3

1. With the above approximation, develop a simple model of the leakage around the turbine section and calculate the flow and lost output due to the leakage.
2. For the following inputs, calculate the present value of the identified seal leakage assuming the second segment of the IP turbine leaks at the same rate as the first:
   (a) Wholesale power price: $50/MWh.
   (b) Operating capacity factor of 96%.
   (c) Discount rate of 11%.
   (d) Period between major overhauls of five years.

1.5.2 Decision Making and Actions – Alternatives

Once a monetary value is determined, the next step in a resolution process would be to identify actions that could be taken to either correct a situation or reasons to delay a decision until a future event makes corrective actions possible. Immediate corrective actions are not warranted every time that issues are identified. In the case of maintenance on a large steam turbine, overhaul expense plus the value of lost production during maintenance easily overwhelm the seal steam-leakage losses. Since the lost output due to seal leakage is minimal, with a relatively small cost, a maintenance decision can be postponed allowing more time to evaluate alternatives and reduce uncertainties.

Reaching a decision to do nothing or postpone a decision requires the same level of involvement and analysis as decisions to take immediate action. Both courses of action are equally as valuable to an owner. If an engineer is employed directly by the owner, or has a position with an outside contractor or consultant, helping the management team reach a decision with clear presentations of results is a value-generating proposition for the engineer.

Though the losses related to steam seal leakage in this case were small compared to the cost of an overhaul, there were several actions that the owner could have pursued to help resolve the uncertainties identified above. First, an in-office review of vibration and differential expansion data from the plant's computer historian could identify if a single event led to contact between the rotating and stationary components. Second a monitoring program could be implemented to observe the difference between measured and expected extraction temperatures or enthalpies between major overhauls.

The third option, or task, would be to review the last outage report. The field service engineer should have recorded "as-found" and "as-left" seal clearances, and these would indicate the quality of the work performed, and condition of the unit following the last overhaul. These findings might indicate that closer supervision of the contractor would be justified. Knowing the cost of the seal leakages could support hiring an outside expert during reassembly of the turbine to help avoid installation problems and future performance losses.

The fourth option would be for the owner to consider advanced designs to improve long-term performance. Again, knowing the value of the losses would help determine if the advance technology parts would be of value and improve the company's returns to shareholders.

1.5.3   Decision Making and Actions – Making a Plan

Knowledge of issues with a thorough analysis of their cost and potential causes is the first step toward improving performance. The next step is to develop a plan, and persuade management that changes in behavior are justified and necessary. At times there are actions within an engineer's control to correct a problem. In most cases, however, resources or approvals will be necessary prior to completing the actions. Even if an individual could easily change a computer program or drawing, change control and QA/QC protocols might require reviews, checks, and approvals. The company's authority limitations, policies, and procedures should be consulted prior to taking individual action. Likewise, within a consulting firm, quality checks and peer reviews need to be completed prior to finalizing a recommendation to the owner or purchaser of services.

Assignment 4

Consider the foregoing and create a short presentation that includes the following related to seal leakage:

- What was found from the testing?
- The value of the lost production due to seal leakage.
- The most probably causes for the leakage.
- Outline a recommended plan to reduce seal-leakage losses in the future.
- Identify resources that would be necessary to carry out the recommended action plan.

1.6   Closure

This case study addresses technical and nontechnical aspects of engineering. On the technical side, seal leakage in an axial flow turbine was observed, corporate engineering standards were improved, and plans were laid for long-term improvements to performance with advanced technology parts.

On the nontechnical side, the case study was a situation that was outside normal practices, giving the engineer an opportunity to demonstrate independence and the ability to work without supervision. As an engineer, take advantage of these opportunities. When you become a supervisor, seek out and reward those who volunteer and who you can depend to work without close supervision. These are the individuals from whom you will learn the most, and who will provide you the opportunity to leverage your position of authority to improve the company's competitiveness.

All rotating equipment that is operated by or operates on a working fluid has leakage between rotating and stationary components. On rare occasions, the leakage can be observed and monitored with common plant instrumentation. This is not always the case, but future developments may provide additional tools to engineers and manufacturers of pumps, compressors, fans, and turbines that could improve their energy efficiency.

Standards and practices are everywhere in engineering. Standards organizations regularly update industry codes; but that is not always the case with practices or guidelines developed within a company. Engineers should be alert to cases in which observations do not necessarily agree with practices in their company manuals. Occasionally take time to research new developments. Read related articles in trade journals and attend conferences sponsored by standards organizations to stay current with your field of engineering. The improvement in corporate standard practices provides new information that could assist with improvements in operating procedures and prevention of future efficiency losses.

When presenting information to management, be clear on the objectives, and use the time you have wisely. If the meeting is for information only, you might expect new assignments on items you have not anticipated. Therefore, think through the possible outcomes before the meeting. If you make a request for help solving a problem or for resources, be specific on the request and show you have a plan to bring the matter to a conclusion. Listen to recommendations and be eager to take on new responsibilities.

1.7   Symbols and Abbreviations

$\dot{m}$:   mass flow
Φ: flow function of a steam turbine stage from equation (1.1)
P: pressure
v: specific volume
"Hga: inches of mercury, absolute pressure

Subscripts

1:   1st stage conditions
C: Corrected
O: observed
t: throttle
d: design or reference
r: reheat intercept valves

1.8   Answer Key

Section 1.1.2

1. (a)
2. (b)
3. (d)

Section 1.3.1, Assignment 1

1. Corrected first-stage pressure = 1865.0 psia (12.859 MPa); +3.04%. Corrected hot-reheat pressure = 600.7 psia (4.142 MPa); −8.27%. Lower throttle flow indicates loss of flow passing ability in the governing stage. The combination of lower throttle flow and an increase in first-stage pressure indicates loss of flow function in the second or subsequent stages of the HP turbine. The hot reheat pressure has fallen more than the change in throttle flow, indicating a slight increase in the flow area of the IP turbine or continued degradation of HP turbine stages beyond the first stage. As the changes occurred gradually over 7 months, deposits in the HP turbine are suspected. A small increase in the flow area of the IP turbine may have occurred.
2. Conditions at the first stage: P = 1868.9 (12.886 MPa) psia, T = 930.4 °F (772.3 K) h = 1434.3 Btu/lb (3336.3 kJ/kg), entropy = 1.539 Btu/lb-R (3.579 kJ/kg-K), specific volume = 0.396 ft^3/lb (0.025 m^3/kg). Conditions at the first stage are solved by trial and error to achieve the desired second HP section turbine efficiency.
3. Flow functions are given in Table 1.3.
4. The conclusions from 1. above are supported by the flow functions – suspected deposits in the HP turbine beginning at the governing stage. Since the change in flow passing ability to the HP turbine second stage is greater than at the throttle, deposits are suspected to be heavier in the later stages of the turbine.

Section 1.3.2, Assignment 2

1. See Figure 1.10. The second extraction measurements show an elevated temperature.
2. Overall IP turbine efficiency: 88.38%.
3. IP section 1: 73.76%, IP section 2: 98.16%.
4. The efficiencies of 3. above are not reasonable. The case study states that the IP efficiency changed very little. A 10% improvement in the second section to near ideal efficiency is unrealistic. If the first section dropped by 15%, the second section would have a similar drop.
5. Expected turbine shell temperature = 887.3 °F (475.2 K). ΔT = 15.1 °F (8.4 K).
6. The most likely cause for the high extraction temperature is blade-tip leakage between rotating and stationary components. Causes for the increased leakage include mechanical damage to the seals caused by high vibration, or differential expansion between the rotor and shell, or an installation error. The seals should be expected to perform

Table 1.3 Turbine flow function results.

	Design customary (SI)	Test customary (SI)	Change (%)
Turbine inlet	41 891 (15.882)	39 254 (14.883)	−6.3
First stage	54 615 (20.707)	49 831 (18.893)	−8.76
Hot reheat	147 878 (56.066)	147 870 (56.062)	−0.01

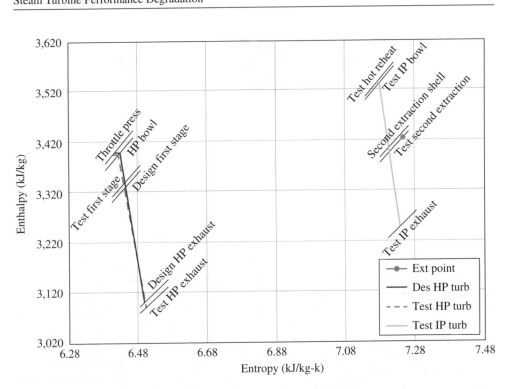

Figure 1.10   Answer Assignment 2, number 1.

reasonably well over a 5-year period between major overhauls supporting the conclusion of mechanical damage or an installation error.

Section 1.5.1, Assignment 3

1. A model similar to the one in Figure 1.11 is expected. Extraction is taken from the periphery of the rotating blades through slots in the turbine casing. Therefore, tip leakage can have a pronounced effect on observed extraction temperature. Leakage flow per section from the model is 20.0 kpph (2.51 kg/s). The lost output from the two sections is 0.72 MW.
2. $1.1 million present value.

Section 1.5.3, Assignment 4

The students should determine if the report is for information or to present a recommendation. If there is a recommendation, it should be accompanied by a request and a plan to implement what has been requested.

The lost output from the seal leakage is far too small to recommend an immediate overhaul. The lost output from the overhaul would overwhelm the 0.72 MW lost from seal leakage. Since something new was discovered that was not included in the company's monitoring program, a recommendation to include a new calculated parameter, shell to extraction temperature difference for example, would be reasonable. An upgrade to better seals would be valuable if the cost of the seals were less than the losses identified, and this too would be a reasonable recommendation.

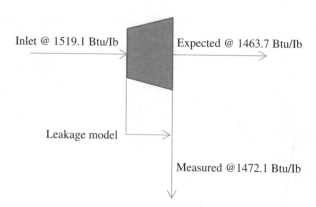

Inlet @ 1519.1 Btu/lb   Expected @ 1463.7 Btu/lb

Leakage model

Measured @ 1472.1 Btu/lb

Figure 1.11   Answer: Assignment 3, number 1.

Improved monitoring of the turbine service provider during maintenance outages would also be a reasonable recommendation. The students may estimate the costs for a consultant and compare that to the value of the lost output as a way to improve their scores on the report. This could require some extra research to determine an appropriate hourly or daily rate to pay the consultant, estimates for travel, and so forth.

The recommendations should be accompanied by a request for resources. In the monitoring case, the engineer can initiate a peer review or similar exercise. Requesting sponsorship from a supervisor would be a helpful addition and could from the basis for the report. A product upgrade suggested by the second outside reading assignment would also be a reasonable recommendation that would require additional resources at the time of the next overhaul. Suggested reviews of actual installations of the newly designed components, which require resources, would be helpful support for a management decision.

An effort to spend additional time studying records, the vibration history for example, should not be included as a recommendation unless additional resources are required. Simple studies should be mentioned in an appendix after the main body of the report, or as something that will be done as a matter of course by the engineer. A good follow-up question would be to inquire why the study has not already been completed. Including results of a fictitious study would be reasonable and could be used to support the student's conclusions.

Following the recommendations, the report should include the steps necessary to bring the matter to closure. Mention of a meeting that has already been scheduled with the turbine manufacturer to discuss product upgrades would show that the engineer has already acted independently and that would improve the value of the report. An outline of the monitoring procedure, as well as supporting cost / benefit calculations can be included in an appendix or as a backup to the main report.

References

ASME PTC Committee (2010) *ASME PTC PM-2010 Performance Monitoring Guidelines for Power Plants*, American Society of Mechanical Engineers, New York, NY.

Cotton, K. C. and Schofield, P. (1970) *Analysis of changes in the performance characteristics of steam turbines. Presented at the ASME Winter Annual Meeting*, New York, NY.

Balling, L. (2010) *Fast Cycling and Rapid Start-up: New Generation of Plants Achieves Impressive Results*, Modern Power Systems, Erlangen, Germany, http://www.energy.siemens.com/hq/pool/hq/power-generation/power-plants/gas-fired-power-plants/combined-cycle-powerplants/Fast_cycling_and_rapid_start-up_US.pdf (accessed January 13, 2016).

Cofer, IV, J. I., Reinker, J. K., Sumner, W. J., *et al.* (1996) *Advances in Steam Path Technology*, GE Power Systems, Schenectady, NY, https://powergen.gepower.com/content/dam/gepower-pgdp/global/en_US/documents/technical/ger/ger-3713e-advances-steam-path-technology.pdf (accessed January 16, 2016).

Pastrana, R. M, Wolfe, C. E., Turnquist, N. A., and Burnett, M. E. (2001) Improved steam turbine leakage control with a brush seal design. Proceedings of the 30th Turbomachinery Symposium, College Station, TX, Texas A&M University, http://turbolab.tamu.edu/proc/turboproc/T30/t30pg033.pdf (accessed January 13, 2016).

Wilcox, M., Baldwin, R., Garcia-Hernandez, A., and Brun, K. (2010) *Guideline for Gas Turbine Inlet air Filtration Systems*, Gas Turbine Research Council, Southwest Research Institute, http://www.gmrc.org/documents/GUIDELINEFORGASTURBINEINLETAIRFILTRATIONSYSTEMS.pdf (accessed January 20, 2016).

Case 2

Risk/Reward Evaluation

Wholesale electricity in Texas is exchanged, bought and sold, and sales are cleared through an open competitive market managed and controlled by the Electric Reliability Council of Texas (ERCOT). Energy prices ($/MWh) are settled in real time or in day-ahead trading for electricity delivered or purchased at nodes throughout the ERCOT system, which is about 85% of the Texas power market. A node is a physical point in the grid where electricity is delivered or consumed. The ERCOT pricing system is an "energy-only" market whereby electricity is sold by the MWh only. The unit MWh is actually a unit of work performed – power (MW) multiplied by time in hours. However, for the purpose of a general understanding in the market place, and for the residents within the ERCOT territory, it is referred to as "energy." The ERCOT territory is approximately 75% of the state of Texas, about 24 million customers, and includes the population centers of Dallas, Houston, Austin, and San Antonio. The electric load managed by ERCOT is about 90% of the load in the state. The actual area covered by ERCOT is shown in Figure 2.1. The ERCOT electric power market is open and transparent with real-time prices available at the ERCOT web site:

http://www.ercot.com/content/cdr/contours/rtmLmpHg.html (accessed January 17, 2016).

Figure 2.1 shows a snap shot of ERCOT wholesale prices ranging from lows of about $35.00/MWh to a high of $125.71/MWh at a location just southwest of Houston.

The ERCOT market allows for electricity to be sold in real time, or to be sold at some future date. Most wholesale power sales happen in a day-ahead market, where a generator will promise and commit to generate a quantity of energy (MWh) at an agreed price for the following day. In general, prices for electricity sold in the day-ahead market run slightly above real-time prices on the actual day of the sale.

Case Studies in Mechanical Engineering: Decision Making, Thermodynamics, Fluid Mechanics and Heat Transfer,
First Edition. Stuart Sabol.
© 2016 John Wiley & Sons, Ltd. Published 2016 by John Wiley & Sons, Ltd.
Companion website: www.wiley.com/go/sabol/mechanical

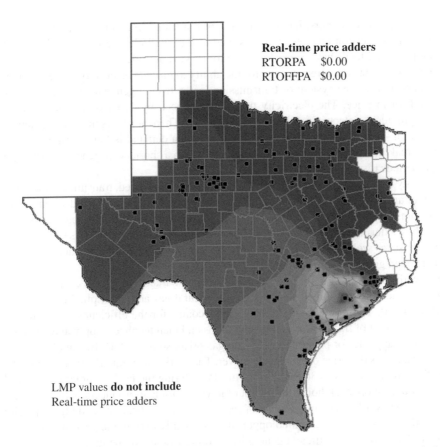

Real-time price adders
RTORPA $0.00
RTOFFPA $0.00

LMP values **do not include**
Real-time price adders

Figure 2.1   Example ERCOT contour map – real time market. *Source*: Reproduced by permission of ERCOT.

Future prices, generally available up to 3 years hence, can follow a similar trend. This is true because the generator/supplier selling power in the future takes a risk that the generating equipment will not be available on the day of sale, whereas a real-time seller does not carry the same risk. The risk premium on the power is often a function of the overall reliability of generating equipment and the time difference between the sale and delivery date. If the generator is unable to provide the energy on the date of the sale, then it must be purchased the market price, and resold at the agreed price to the original purchaser. As demonstrated by Figure 2.1, the real-time market can present a substantial risk to purchases or real-time power.

Long-term sales of up to 10 years or more, or sales with different terms, may be arranged through bilateral agreements, the terms of which are determined by the seller and purchaser. For example, a generator may wish to sell power with a capacity mechanism ($/MW) to guarantee payments that cover the cost of capital to build, as a call option, or a forward energy-only price as a hedge against price volatility. Bilateral agreements are made outside the open ERCOT market, and their terms are usually private. However, the ERCOT market provides a benchmark for the price of wholesale power.

Retail prices to consumers, include the cost of transmission and distribution, fees for the type of load, congestion charges, and fees paid to the retail electric provider. Large, nearly constant industrial loads have lower fees than for service provided to a residence.

The nodal market structure accounts for the ability to provide electricity at the point of use. Thus, congestion on the system or transmission losses from a generator to a node would be accounted for in price. The electricity price is influenced by demand and the supply – the ability to provide the electricity when it is sold or used. There are cases in which an abundance of generation – for example, from wind at a time when demand is low – results in negative prices for real time power. There are also cases in which a shortage of generation during peak demand periods results in exceptionally high pricing.

Electricity is the only consumer commodity that is manufactured, transmitted, metered, and used simultaneously; it cannot be stored. Thus electricity must be generated when demanded by a customer and ERCOT is, in turn, obligated to regulate the maximum legal price to avoid market abuses during times of shortages. The maximum price that ERCOT sets is intended to prevent abuse, as well as to provide incentive for generators to build new capability at nodes where demand growth outstrips supply.

On any given day, the ERCOT power price is benchmarked against the cost of natural gas. Electricity prices are computed with the cost of natural gas and an implied market heat rate (MMBtu/MWh). The implied market heat rate accounts for the efficiency in manufacturing the power, the cost of fixed and variable operations and maintenance, supply and demand, and a generator's appetite for profit and risk. The gas price, sold in $/MMBtu, multiplied by the market heat rate is the price of wholesale power. The market heat rate fluctuates with demand normally ranging as high as 30 to 40 MMBtu/MWh during the peak summer months to lows of about 4 during off-peak hours throughout the year. Efficient producers with low operating heat rates are rewarded with more hours of sales at the market heat rate and higher profit margins. Inefficient players have limited opportunities, usually only in the peak summer months, or are driven to build new efficient generation in order to stay competitive.

While there are unique aspects of the ERCOT electricity market, it shares supply and demand principles with open markets regardless of the commodity bought and sold. Efficient, low-cost suppliers and those that can quickly take advantage of market pressures are more successful and the consumers are provided with low-cost products.

2.1   Case Study

An independent power producer owns a single 478 MW combined-cycle gas turbine (CCGT) generating facility near a large population center. There are several large power producers in the vicinity capable of supplying the local demand. The CCGT facility consists of two identical natural gas-fired gas turbines, each exhausting into a horizontal flow heat recovery steam generator (HRSG) that provides steam to a common steam turbine ($2 \times 2 \times 1$ configuration). The full load operating HHV efficiency of the facility is 47.89%.

The owner has had a few tough years with profits below expectations, and employees are worried about the continued viability of the small company. A large conglomerate may purchase the company, and, if so, their jobs could be in jeopardy.

Unexpectedly, an unplanned outage over the weekend at a large nearby generating facility has caused a sudden shortage of generation at its ERCOT node. Real-time prices on Monday morning soared to near the legal limit, and the company's power traders anticipate that day-ahead prices will remain at least three times the normal value for the next 3 weeks while repairs are completed, and the nearby facility is returned to normal service.

Company managers have requested an engineering assessment of whether they should delay the current scheduled hot gas path inspection (HGPI) on one of the two gas turbines and continue to generate through the anticipated high-priced period of the next 3 weeks. The outage is schedule to begin Monday evening after the peak usage period. To reduce operating risks and stress on the unit, it would be operated throughout the period without an outage.

Prior to the incident, wholesale prices were averaging $41.34/MWh during the 16-hour peak period of hours from 06:00 to 22:00. During the 3-week period, off-peak prices are anticipated to be unchanged from normal.

The situation is complicated by the fact that the gas turbine scheduled for maintenance has operated through its 24 000 hour maintenance interval plus an 800 hour allowance offered by the manufacturer without endangering the manufacturer parts warranty. Continued operation would place any risk of failure on the owner with all costs for repairs and collateral damage suffered by the owner. In addition, the labor force for the maintenance overhaul has been marshalled and is currently completing safety training for site access, which will be completed by the end of the day. Any further delay of the outage start date could jeopardize the skilled labor force availability, resulting in additional costs and, perhaps, the loss of key personnel that have proven capabilities to provide quality workmanship, and on-time return to service.

In addition, a delay to take advantage of the near-term high power prices would place the outage in a period of the summer time peak prices making the decision more difficult. Management is weighing options of operating with certain high prices now compared to uncertain conditions later in the summer. The engineering task is to determine if the risks associated with continued operation outweigh the gains of the short-term market opportunity.

2.2   Background

2.2.1   Types of Gas Turbine Generating Plants

Operating characteristics, and relatively high exhaust temperature of gas turbines afford a wide variety of applications in the generation, and co-generation of power and steam. The most direct application is in simple-cycle, or single-cycle (SC) power generation, using the fast-starting capability of gas turbines to provide near-term supplemental power to the electric grid. These applications are often called "peaking plants" or simply "peakers" as they are primarily employed during the peak load periods of the day, or when there is a sudden increase in power price. Peaking plants are often used to back up renewable energy sources that depend on weather or time of day.

Gas turbines in peaking services are mostly modified gas-turbine models designed for aircraft: aeroderivatives. Aeroderivatives are lightweight, with short starting times, and typically high compression ratios that yield lower exhaust temperatures than larger frame machines for utility service.

Figure 2.2 shows a sketch of a gas turbine exhausting into an HRSG, abbreviated "1 × 1." This arrangement is often used in combined heat and power (CHP) facilities to provide the host company or facility with a portion of its steam requirements. A drawback of the simple configuration shown in Figure 2.2 is that the steam generation is a function of the gas turbine load. A host with a variable steam demand with a majority of the steam provided by the CHP would result in unpredictable power sales. An improvement to the design is to add

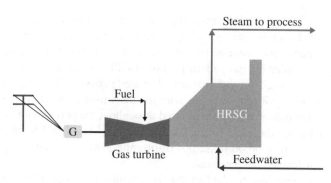

Figure 2.2   1 × 1 CHP.

supplemental firing in the HRSG, sometimes referred to as "duct firing" to provide the flexi-bility require by the host.

Another drawback of this design is the reliability of the steam supply. Where steam is critical to the host facility, a second gas turbine/HRSG combination, backup steam supply, or both may be required to satisfy reliability requirements.

The net effective efficiency for this type of cycle can be very high because the latent heat of condensation is treated as useful energy supplied to the process host. Combining the energy outputs of electricity and heat can yield efficiency well over 50% on a higher heating value basis.

An HRSG is a large heat exchanger converting the energy in a high-temperature exhaust gas, usually a gas turbine exhaust into steam. Figure 2.3 is a three-dimensional rendering of a three-pressure subcritical HRSG. The gas turbine exhaust enters the HRSG from the left through the diffusion section. Across the top are the HP, IP and LP steam drums in the direction of steam flow. An emissions removal catalyst section is shown in the drawing following the HP evaporator section. The LP section is an integral deaerator, which provides removal of essentially all the oxygen in the feedwater to the remainder of the steam cycle. Exhaust gases exit through the stack located on the right in the figure. A photograph of a similar unit is shown in Figure 2.4.

Addition of a steam turbine to create a combined cycle (Brayton gas turbine cycle plus a Rankine steam cycle,) as shown in Figure 2.5, improves the efficiency of power generation. Here, a nonreheat steam cycle operating on the waste heat from a gas turbine adds additional electrical output from the single fuel source creating a 1 × 1 ×1 configuration (one gas tur-bine, one HRSG, and one steam turbine.) More complex steam cycles including reheat can further improve the overall cycle efficiency through better use of the gas turbine waste heat. The steam turbine output in these cycles is nominally one-third of the gas turbine generator output. Due to the rejection of heat in the condenser, these power cycles have efficiencies generally lower than well designed simple cycle CHP facilities consisting of a gas turbine and HRSG.

As with a 1 × 1 design, duct firing can be added to the 1 × 1 ×1 to provide the ability to generate more electricity in the steam turbine when market conditions are favorable. This additional capacity is an economic choice built at a cost ranging from one-third to one-half the cost of the base facility on a capacity basis of $/MW. But it comes with an efficiency cost. When duct firing is not used, the steam turbine operates a nonoptimum output thereby reducing

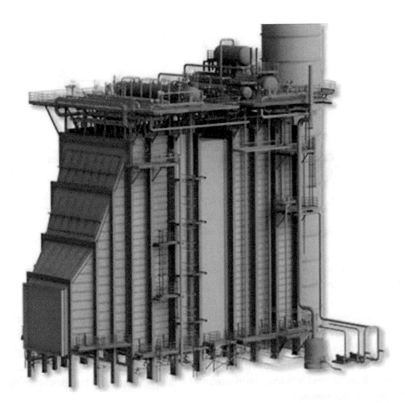

Figure 2.3 Rendering of HRSG. *Source*: Reproduced by permission of Nooter/Eriksen.

Figure 2.4 A photograph of an installed HRSG. *Source*: Reproduced by permission of Nooter/Eriksen.

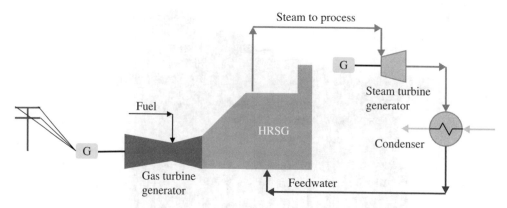

Figure 2.5   1 × 1 × 1 combined cycle generating plant.

the base cycle efficiency. When duct firing is in service, generation is added incrementally at a single Rankine cycle efficiency lowering the overall combined cycle efficiency.

A common configuration for combined cycle generating plants is a 2 × 2 × 1 – two gas turbines, two HRSGs and one steam turbine. This configuration improves the economy of scale by reducing the steam cycle costs in proportion to those of the gas turbine/HRSG combination. Unlike gas turbines, which are manufactured in discrete sizes and models, steam turbines are custom built for the specific application. Therefore, the steam cycle can be optimized for one or more gas turbines, environmental constraints and the business case to improve the overall project economics.

Combined cycle CHP plants are a combination of Figure 2.2 and Figure 2.3. Process steam may be provided directly from the HRSG or through an extraction port in the steam turbine. The efficiency of these CHP facilities can approach that of the SC CHP but will always be less due to the condenser heat rejection. Combined cycle CHP facilities provide excellent operating flexibility for steam and power supply when combined with duct firing. Electrical output can be maintained at a constant level through variations in process steam demand. The process steam demand may be provided from a variety of sources – direct from the HRSG or extracted from the steam turbine. Adding gas turbines, and steam turbines to the facility design improves reliability offering the ability to provide steam during normal or unexpected maintenance outages.

Auxiliary features can be added to the gas turbine cycle including:

- Steam power augmentation: steam injection downstream of the combustion zone to increase output.
- Water or Steam injection for NOx control: water or steam injected in the combustion zone to moderate the flame temperature and reduce the production of NOx.
- Inlet air cooling to increase mass flow and gas turbine output in the summer.
  - Inlet fogging: direct sprays of water into the air inlet duct to the gas turbine.
  - Inlet evaporative cooling: indirect evaporative cooling.
  - Mechanical inlet air chilling: using a refrigeration cycle to chill the inlet air stream.
- Compressor water injection: water spray at an intermediate point within the compressor section to lower the compressor temperature and add mass flow to the latter stages of the gas turbine.

- Wet compression: a combination of inlet fogging and compressor water injection carried out in the inlet duct. A quantity of water in excess of that required to reach 100% relative humidity sprays into the inlet air stream resulting in liquid droplets of water entering the first stages of the gas turbine. These droplets evaporate in latter compression stages, providing cooling and reducing power required for compression.

2.3   Gas Turbine Operating Risks

Large-frame advanced technology gas turbines such as the one shown in Figure 2.6 operate with firing temperatures well above the melting point of superalloys used in the manufacture the combustion zone parts and the first several rows of the turbine blades and vanes.

Figure 2.6   Gas turbine cut away. *Source*: Reproduced by permission of General Electric.

To maintain component integrity, a thermal barrier coating (TBC) is applied to the hot gas path components in the hottest regions of the machine to insulate the base metal from the gases at the firing temperature. Additionally, air from the gas compressor or, in some cases, steam flows through internal cooling passages of the blades and vanes to maintain metal temperatures within an acceptable range. If air is used, it exhausts through vents in the blades and enters the hot gas path. When steam cooling is used in advanced technology gas turbines, it is routed from the blade passages to the HRSG steam cycle to generate additional power in a steam turbine.

The TBC generally comprises several layers with the top layer containing a ceramic oxide, often zirconium oxide. A manufacturer will conduct extensive laboratory and field testing of its formulation to establish a life expectancy and maintenance intervals for the specific gas-turbine model. While failure mechanisms are not completely understood, coatings generally exhibit cracking and spalling from the surface toward the end of life, exposing the base metal to high-temperature gases, which may lead to burning away of a base metal and, in some cases, liberation of parts in the combustion zone, a blade or vane.

Failures inside a gas turbine can be fantastic. Consequential damage following the initial failure may include the compressor, combustion section, turbine blades and vanes, the exhaust diffuser, and the casings of the compressor and turbine. Figure 2.7 shows the condition of the rotating blades of the gas compressor after a liberation of a second-row (R-1) blade. The owner replaced all rotating and stationary blades of the compressor section, in addition to combustion parts, and the first two rows of the turbine damaged by molten metal and shrapnel from the combustor following the compressor blade failure. Rather than reblade a new compressor rotor, a spare rotor was purchased and flown in from across the globe to reduce lost production.

Figure 2.7 Compressor consequential blade damage. *Source*: Reproduced by permission of General Electric.

Figure 2.8 Failed turbine wheel and consequential damage. *Source*: Reproduced by permission of Phillips 66.

Figure 2.8 shows a failed section of the turbine first stage wheel that held seven blades. The lower left-hand corner of the photograph shows where first-row blades once were. The air-cooling holes are visible on the blade base plates. The first-stage vanes are visibly damaged by splatter, as is the second rotating blade row due to shrapnel passing through the operating unit. Damage to the gas turbine resulting from this failure required replacement of all blades and vanes in the compressor and turbine sections, the compressor and turbine rotors, all of the combustion hardware, most of the compressor casing sections, the complete turbine casing, and half of the exhaust diffuser. The inlet bell housing had minor damage that was repairable. The repair cost was more than the price of the original machine. The component that failed was outside its warranty period, and therefore not covered by the manufacturer. Consequential damage repair was therefore the responsibility of the owner.

2.3.1 Gas Turbine Major Maintenance

To guard against catastrophic failures, gas-turbine manufacturers recommend a program of major maintenance based on the operation and technologies used for the components inside the machine. For a typical large-frame gas turbine, inspection intervals for the combustion section, hot gas path, and major inspections are specified.

The combustion section, containing the hardware that combines the fuel with air from the compressor section and provides combustion gas to the turbine section generally has the shortest operating life of the major sections of the gas turbine. These parts include the fuel nozzles, combustion cans where the fuel burns, and transition pieces that connect the combustor cans to the turbine section inlet vanes. Combustion inspections (CI) may be recommended every 8000 hours for some gas-turbine models and all parts within the combustion section are removed and replaced with new or refurbished parts.

A hot gas path Inspection (HGPI) encompasses the combustion section as well as the turbine section of the gas turbine. During this outage, parts in the combustion section would be replaced as in the case of the combustion inspection. Blades and vanes of the turbine section may also be replaced due to damage suffered during the maintenance interval, or based on the equivalent operating time accumulated. Rows of blades and vanes closer to the combustion section have a shorter expected life than the exhaust blade row, due to the operating temperature.

The HGPI will also include disassembly of the bearings supporting the turbine section rotor. The bearings will be sent to a specialty maintenance facility to reapply the Babbitt. The lubricating oil may also be inspected, and replaced if necessary.

A major inspection is a full disassembly of the gas turbine internal rotating and stationary parts. The inspection includes a CI and HGPI, as well as an inspection of the compressor section. As with the CI and HGPI, components are replaced as described above. Compressor parts have longer expected lives than the parts exposed to the hot combustion gases; so, maintenance and replacements are less frequent. Compressor blades and vanes may have a design life as long as the rotor.

Rotor-life extension or rotor-life assessment inspections are the least frequent of all gas turbine inspections. The first assessment of the rotor life may be after 12 years of continuous operation. Typically, manufacturers do not guarantee a rotor life. They will specify the expected rotor lifetime and can inspect the rotor after that period to see how well the rotor has maintained its integrity. Parts that are subject to the highest stresses, or the most severe duty, can be inspected, or the entire rotor may be disassembles and thoroughly inspected. Those parts that show signs of failure are usually replaced. The owner may choose to inspect all the components or simply replace the rotors at the end of the design life depending on their risk tolerance.

Without a full rotor replacement, most manufacturers will not recommend the rotor for further service. However, a gas turbine rotor is very expensive, and industry experience can be a guide to decision making.

Labor for gas turbine major maintenance is specialized. Craft labor specifically trained to perform the maintenance as well as the supervision are required to be at the owner's facility prior to and during the actual outage. Planning for the outage, which includes identification of the labor, materials, parts, and offsite maintenance facilities can begin as much as 6 months in advance of an outage. The client and supplier work closely together to define the scope and timing of the outage, safety and training requirements specific to the site, and so forth. Additional work that develops during the outage due to the inspection nature of the maintenance can add significantly to the costs and duration of the outage. Operating history that could imply extra work would be thoroughly reviewed and analyzed during the planning period to reduce risks during the outage.

2.3.2   Equivalent Fired Hours

Manufacturers' maintenance intervals for gas turbines are based on equivalent running hours, or equivalent fired hours of operation. A turbine operating for an hour at full load under steady-state conditions is typically measured as one equivalent hour. Starting and stopping add equivalent hours due the thermal stress imposed on the hot components. A fast start, or trip from full load, add additional hours, as do operations at peak conditions, those above the full load guarantee, changes to liquid fuel, steam addition for power augmentation, and so forth. A manufacturer will provide a methodology and equations to determine the equivalent fired hours in the operating manual or in a contract for major maintenance.

After laboratory testing, and extensive field experience, manufacturers are quite certain of the mean time to failure for their TBC and each part within the gas turbine. Based on the life expectancy, they provide a conservative maintenance interval that is short enough to prevent most failures but long enough to remain competitive in a global market. Since outage times may depend on seasonal pricing, the availability of a skilled workforce, and other considerations, owners generally have a buffer around the recommended interval to provide some flexibility in scheduling a gas turbine maintenance outage. Beyond the buffer period, in this case 800 hours, the manufacturer will not accept liability for warranty coverage, leaving the owner with a substantial risk. The owner could be responsible for all costs of repair less any allowance for a normal major maintenance inspection.

2.3.3   Failure Costs

After accounting for a regularly scheduled maintenance outage, repair costs from a failure could be as high as $10 to $30 million on a large-frame gas turbine. Furthermore, parts included in the collateral damage of the failure may not be immediately available, resulting in an outage lasting 12 to 18 weeks instead of a planned outage that may only be 10 days.

Repair and lost production costs would be just the beginning of higher operating costs for an owner following a failure. Insurance as well as borrowing costs can increase. The owner may find it difficult or impossible to obtain insurance, which could place the owner in default of loan covenants requiring payment in full of any outstanding debt. The only good news would be that the unit could be repaired to nearly its original condition.

2.3.4   Reading Assignment

Read Clark *et al.* (2012).

1. What portion of the world's electricity production is from gas turbines?
    (a) 10%
    (b) 20%
    (c) 30%
    (d) 35%
2. Reasons to improve TBCs include:
    (a) Improved gas turbine efficiency.
    (b) Greater thrust-to-weight ratio.
    (c) Higher durability.
    (d) All of the above.
3. Functions of the TBC include:
    (a) Thermal insulation.
    (b) Reflective to radiant heat from combustion.
    (c) Strain compliance during thermal cycles.
    (d) All of the above.

2.4   Case Study Evaluations

2.4.1   Review

This case study weighs the benefits of operating during a period of higher prices against the potential risks of continued operation. The gas turbine in question has exhausted its maintenance interval plus the maintenance interval buffer allowed by the manufacturer. The potential gains and losses are substantial. The company performance over the past few years suggests a need to improve revenues and margins.

There are a couple of approaches to solving the problem. A common approach would be to build a risk tree. On the tree will be the various decision opportunities, their potential gains or losses, and probability. Multiplying each gain or loss by its probability and summing the products shows the most probably outcome – the expected value presented by the current market conditions. One possible decision tree is shown in Figure 2.9.

Completing a decision tree often requires employing the opinions of subject-matter experts (SMEs), and working through various techniques to develop the event probabilities and calculate the potential gain or loss from each possible branch of the tree. In a small company, the engineer tasked to perform a task may be the only subject matter expert. In a large organization, tracking down the SMEs for a time-constrained decision process can be impractical.

Another option is to calculate the probability and outcome for a single branch of a decision tree, or even a partial branch, then calculate the sensitivity of the outcome based on the input assumptions. This technique does not produce an expected value based on the probability of the outcomes, but instead produces a range over which the outcome may be expected to lie. This technique can provide mangers with the data necessary to make an informed decision by demonstrating how the boundaries of assumed values affect the outcome.

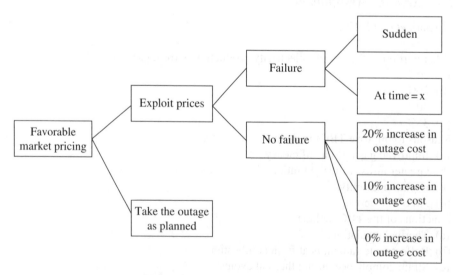

Figure 2.9   Possible decision tree.

2.4.2   Presenting Results

This case is also about developing and presenting results quickly to avoid being overcome by events that would diminish the value of the work or make the decision for you. In such cases, having a few generic presentation tools readily available can improve the chances of getting the message across to the right people in the time available. For this case study, a "tornado" diagram was chosen to capture the range of uncertainties with a simple presentation. This type of stacked bar chart gets its name from its shape. Items with the greatest uncertainty – the largest positive and negative variances from a base case – are shown at the top while those with the least impact on the decision are at the bottom. The resulting inverted triangle shape focuses attention on the items with the greatest importance for the results, and helps drive a decision based on the certainty of information that is available at the time.

To construct a tornado bar chart from sensitivity studies in Excel, use a "stacked bar chart" with two series, lower and higher earnings or returns, for each sensitivity case. Sort the cases in order of increasing absolute value using either the lower or higher series, to provide clarity and understanding for the audience. Figure 2.10 is an example Tornado diagram using the data shown in Table 2.1. In this example, Uncertainty E would be the focus of any discussion, with a possible mention of Uncertainty D. Notice also that Uncertainty E has a greater positive impact than negative. A strategy to help a project or decision based on this type of result could focus on the methods to ensure a positive outcome from Uncertainty E rather than passively accepting its potential.

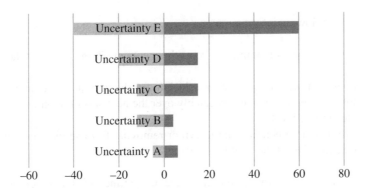

Figure 2.10   Example tornado diagram.

Table 2.1   Example data for tornado diagram.

Study case	Lower earnings	Higher earnings
Uncertainty A	−5	6
Uncertainty B	−12	4
Uncertainty C	−12	15
Uncertainty D	−20	15
Uncertainty E	−40	60

The critical time factor of this case study will govern how the results are presented. Developing a formal PowerPoint slide presentation is probably not the best use of time. Instead, a quick e-mail with the conclusion and a tornado diagram would be more effective. A phone call can be suitable but should be followed with an e-mail to document the conversation and any decision.

2.4.3   Judgment Calls

The following assumptions help begin the evaluations and will be tested to determine their significance on the decision outcome.

1. Let the standard deviation (σ) around the mean time between TBC failures ("MTBF") be 5% of the recommended maintenance interval – 1200 hours.
2. Assume there are two methods of arriving at the maintenance interval (MI):
    (a) For the base case, assume the manufacturer has a small margin (δ) subtracted from the actual mean time to failure less two times the standard deviation (σ). That is, for Method 1: $MI_1 = MTBF - \delta - 2*\sigma$. Assume the margin ($\delta$) yields a cumulative probability of a TBC failure of 1% at the recommended maintenance interval.
    (b) In the alternative case, the manufacturer sets the maintenance interval at the actual mean time to failure less three times the standard deviation. The maintenance interval for Method 2 is: $\left(MI = MTBF - 3*\sigma \right)$.
3. A catastrophic event would cost the company $25 million in addition to the planned HGPI outage.

You also have the following information:

- A standard HGPI inspection nominally costs the company $5 million and takes 3 weeks to complete.
- Natural gas has been purchased on a set schedule for the current month at $4.50/MMBtu. Prices are not expected to change appreciably over the next several months.
- The corporate tax rate is 28%.
- Accounting for cycling operation (turndown or removing from service on a nightly basis) the capacity factor (CF) of the facility is 65%.

From the data and base case assumptions, the actual mean time to failure is equal to the recommended maintenance interval plus two times σ, plus a small margin.

2.4.4   Exercise

1. Use the Excel NORMDIST function to determine the manufacturer's margin such that the cumulative probability of failure at the recommended maintenance interval is ~1% from 2(a) above.
2. Calculate the probability of failure at the recommended interval plus 800 hours.
3. Find the probability of failure at the end of the proposed 3 week run-time extension.
4. Calculate the normal marginal cost of production ($/MWh) neglecting nonfuel operating and maintenance costs.
5. Calculate the spark spread ($/MWh) at the normal and temporary market power price for the 16-hour daily peak.

6. Determine the probable lost profit opportunity that would occur from a 3-month outage extension. The loss would include gross margin plus the differential maintenance cost due to a catastrophic failure that could occur within the 21-day operating extension.
7. Calculate the base case probable gain or loss for extending the run time 21 days for the base case assumptions.

2.4.5 Sensitivities

Once a value for the base case has been determined, testing of the assumptions helps provide credibility or provides guidance to change the basis and arrive at a new base case. For this case study, there are a number of assumptions that should be tested, including:

- Engineering assumptions:
  - The value of σ for the two methods of determining the maintenance interval – MI.
  - The manufacturer's margin – δ.
  - The assumed method of setting the recommended maintenance interval (2σ + margin "Method 1" versus 3σ "Method 2").
- Commercial assumptions:
  - The duration that prices will remain at current levels – three times the normal rate.
  - The actual price that will exist during the assumed period of three weeks.

2.4.6 Exercise – Sensitivities

Even though management requested an engineer's opinion, the engineer should not forget the commercial side of the equation. Commercial assumptions on price and quantity are often much less certain than engineering assumptions and can carry substantial impact on a decision. For the above parameters, check the net value of the extended run time for changes of 50% (worse) and 20% (better). That is:

- Method 1 σ: −50%, +20%.
- Manufacturer's margin: −50%, +20%.
- Power price: −50%, +20%.
- Temporary increase period: −50%, +20%.
- Method 2: 3σ with no margin.
- Method 2 σ: −50%, +20%.

2.4.7 Presentation of Results

Use a stacked bar chart in Excel to prepare a "tornado" diagram. To do so, there will be two data series: the results with a positive impact over the base case, and those with negative impact. The axis categories are:

- Manufacturer margin.
- Method 1 σ.
- Method 2 3σ no Margin.
- Method 2 σ.
- Duration of high prices.
- Price.

There are positive and negative impact values for each category listed, some may be zero. Assign colors for the two series to help convey the monetary meaning of the results. Plot the variances from the base case and provide appropriate labels for understanding.

Prepare an e-mail with your conclusion and recommendation whether to take the outage immediately or delay for a specific period.

2.5 Case Study Results

Continuing with the scheduled HGPI overhaul outage as planned was supported by engineering analysis showing a possible profit that could easily be overwhelmed by technical and commercial uncertainties. Management chose the conservative alternative rather than take a greater than 10% chance of a major catastrophic failure. The outage started as planned on Monday evening. By the time the unit was cooled and owner responsibilities complete, contractor personnel were on site and the outage was concluded on schedule.

During the condition assessment, after the turbine casing was removed, evidence of minor domestic object damage (DOD) was apparent on the first-stage turbine blades. A part from the combustion section, most likely a bolt, released and knocked a small chip of TBC from the center of a first stage blade. The manufacturer's site engineer provided guidance that the chip was small enough so that the machine could operate until the next major inspection in about 3 years without replacing the blade. Bore scope inspections of the blade's condition would take place once or twice per year over weekend outages to ensure the minor damage did not progress into a more serious situation.

The evidence of DOD and the inspection of the combustion section led to the conclusion that further combustion section damage was a few short hours away. The bolt that had released resulted in leaks in the combustor section that were leading to a rapid decline in the integrity of a combustor can. Continued operation for an additional 21 days would certainly have led to catastrophic failure of a combustor can with consequential damage that could have included the first two rows of the turbine. The company acknowledged a near miss of a serious accident. From this experience and the improved understanding of the growing risks over time, future outages were scheduled close to the manufacturer's recommended maintenance intervals. Management approvals were implemented for schedule changes that encroached into the 800 hour buffer.

Commercially, prices were strong on Tuesday; but dropped to near normal values by Wednesday. Other generators in the vicinity noticed the temporary shortage and were able to manage a slight increase in profits by increasing generation at slightly increased wholesale prices at the node. The short-term spike, with the accompanied high emotions, quickly subsided back to a more normal situation.

2.6 Closure

Students should take away the following concepts from this case study:

1. Even when faced with very little information, good engineering judgment together with tests of assumed values can be enough for sound decisions.
2. Build and maintain a "tool box" of presentation methods that help convey complex ideas in simple to understand ways.
3. Do not neglect the commercial considerations that affect decisions.

4. Be aware of contract provisions that affect the work you are doing. This applies to every engineering career including consulting engineers, manufacturers, service providers, designers, and owners.

2.7 Answer Key

External reading

1. (b)
2. (d)
3. (d)

Section 2.4.4

1. Manufacturer's margin 1.3% of the MTBF including margin or 358 hours. See Figure 2.11. This value is determined by trial and error. Use of the "goal seek" function within Excel is useful.
2. Failure probability at interval + 800 hours: 5.1% (=NORMDIST(MI + 800,MTBF,σ,TRUE).
3. Failure probability at end of a 3-week extension: 11.3% (NORMDIST(MI + 800 + 21*24,MTBF,σ,TRUE).
4. Marginal cost: $32.06/MWh. Gas price ($/MMBtu) * HR_a (MMBtu/MWh) where: HR_a = actual heat rate (3.41214/cycle efficiency).
5. Before tax spark spread: Normal $9.28/MWh, temporary: $91.96/MWh. Spark spread = gas price (MMBtu/MWh) * (HR_M − HR_a), where HR_M = market heat rate = market price $/MWh/gas price $/MMBtu.

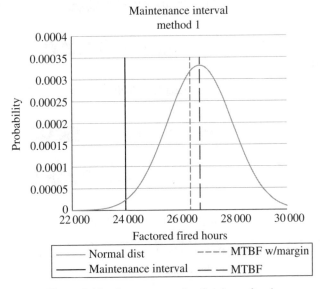

Figure 2.11   Answer to section 2.4.4, number 1.

6. Three-month outage lost profit opportunity at 65% capacity factor: $2.29 million. LPO = { normal spark spread * capacity * 3 months * Days/month * 24 hours/day * CF + catastrophic repair cost } * failure probability * (1 − tax rate). Capacity = total plant capacity/2.
7. Base case gain for 3-week extension: $2.5 million. Possible gain − LPO. Possible gain = (temporary − normal spark spread) * capacity * 21 days * 16 hours/day * (1 − tax rate) = $4.78 million.

Section 2.4.5

Repeat items 1–7 above for the sensitivity cases. Sort smallest to largest and plot on a stacked bar chart. Select bar colors that clearly represent losses or gains.
Repeat items 1–7 above for the sensitivity cases.

The results above assume the manufacturer's margin is recalculated for the −50% and +20% standard deviation cases of Method 1. The sensitivity results (Figure 2.12) show that

Table 2.2   Answer – section 2.4.5, sensitivities.

Summary	Lower	Higher
Mfg. Margin −20% +50%	−$637 105	$226 891
Method 2 3σ	$0	$1 721 605
Method 2 σ −50% + 20%	−$1 853 463	$1 919 098
Duration of increase	−$2 048 771	$0
Power price	−$3 585 349	$1 434 139
Method 1 σ −50% +20%	−$6 638 200	$692 762

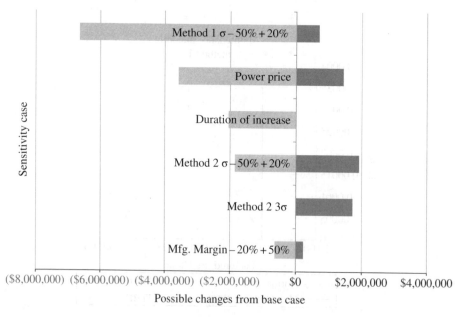

Figure 2.12   Answer – section 2.4.5, tornado diagram.

the engineering assumptions can be significant and would change the results substantially. The downside risks from technical and commercial assumptions could easily overwhelm the expected base-case gain of only $2.5 million. Taking into account the other downside risks associated with the labor force, there are several factors that could lead to a decision to take the HGPI outage as planned.

Notwithstanding, there are cogent arguments for continuing operations including:

- the expected revenue increase;
- risks of a failure in the range of 10% are reasonable;
- there are upside potential gains over the base-case assumptions;
- improved corporate earnings could improve morale.

Reference

Clark, D. R., Oechsner, M., and Padture, N. P. (2012) Thermal-barrier coatings for more efficient gas-turbine engines. *MRS Bulletin* **37**(10), 891–898.

Case 3

Gas Turbine Compressor Fouling

An independent power producer (IPP) owns and operates natural combined cycle gas-fired gas turbine (CCGT) power plants throughout the United States and in the United Kingdom. Several are combined heat and power (CHP) plants collocated with a chemical plant or refinery, and sell energy output as steam to the host industrial plant in addition to electricity to the host and to the local market.

Combined heat and power plants recover a portion of their fuel costs through steam sales to the host facility, which improves the effective efficiency of power generation. The energy value of steam sold for process heat includes the latent heat of condensation and cooling to the site ambient temperature, which improves the overall cogeneration plant economics and gives the CHP operator an advantage in the power market. The host facilities enjoy a negotiated price of steam that is less than they are able to self-generate, and no direct emissions are attributed to the host for the energy supply. The savings in emissions attributed to the host can facilitate future capital expansion projects.

Reliability of power supplied to the industrial plants by the CHP plant is assured by a backup provided by the local electric grid. Contract requirements often require additional steam-generating equipment to be installed at the CHP to ensure near 100% steam reliability.

Contract agreements between the host and cogenerator should account for the additional capital required to build redundant steam supplies and the captive nature of the host/cogenerator relationship. In most cases, a cogenerator would be required to operate, even generating power at a loss, to maintain the steam supply. Therefore, the prices for electricity and steam often account for the host's "capture" of the cogenerator.

Electricity prices have become more competitive worldwide. Newer power plants with advanced technology gas turbines, government subsidies, and special tax treatments for renewable energy have cut into the IPP's margins and there is now concern over maintaining

Case Studies in Mechanical Engineering: Decision Making, Thermodynamics, Fluid Mechanics and Heat Transfer, First Edition. Stuart Sabol.
© 2016 John Wiley & Sons, Ltd. Published 2016 by John Wiley & Sons, Ltd.
Companion website: www.wiley.com/go/sabol/mechanical

shareholder returns. In the United Kingdom, an absence of abundant natural gas has forced importation of high-cost LNG, resulting in elevated fuel prices placing further pressure on profit margins.

Cost-cutting measures by the IPP to improve returns have reduced their engineering capability and they now mostly rely on outside engineering consultants to assist them with various studies and capital improvement projects. Your engineering consulting company has been commissioned to review the IPP's maintenance practices and focus on their largest facilities in an effort to improve reliability and efficiency.

From your company's initial review, the IPP has a variety of gas turbine models in its portfolio, which were provided by leading industrial suppliers. They follow manufacturer recommendations and industry best practices for maintenance intervals. Their operators are well trained, and the staffing level at each facility is consistent with good industry practice. Their safety records show steady improvement and they are in the top quartile in their industry regarding accident rates.

Given their processes and overall performance, your company believes that new thinking would be required to make a noticeable effect on the IPP's bottom line. New technologies, or new methods not recognized by the manufacturers at the time the units were purchased may be necessary, and those could require manufacturer concurrence prior to implementation to avoid loss of warranty coverage.

Your review of outage times at one of the major cogeneration facilities showed potential for such a change. This facility, located in a heavy industrial area near the ocean, behaves in a similar way to most gas turbine plants and suffers from steady compressor fouling from airborne dust and water vapor. Each day, the operators conduct an online water wash on each of three gas turbines by spraying water into the inlet of the gas compressor. In addition, the machines are regularly removed from service to conduct an offline, crank-soak wash, to remove deposits and help return the units to a clean condition.

During the 48-hour offline washes, auxiliary boiler operation is required to assure a backup steam supply to the host refinery. The backup steam generation is less efficient than the use of waste heat from the gas-turbine exhaust; therefore, the IPP experiences higher fuel expenses in addition to lost generating opportunities. The outages are conducted on the weekends, so they also suffer from higher labor costs to conduct the maintenance.

Your research into this practice was a key element in your company's recommendation for the IPP to optimize the maintenance cycle and consider replacement of the gas turbine inlet filtration systems to completely eliminate the need for water washing. If successful in persuading the IPP to make the modifications with your company, new engineering, construction management, and purchasing resources would be required, representing a significant growth opportunity for you and your company.

3.1 Background

3.1.1 Gas Turbine Types

Gas turbines used in power generation fall broadly into two categories: aero-derivative and frame machines. Aero-derivative units are modified aircraft machines with a power turbine added to drive a synchronous generator. Aero-derivative units typically have higher compression ratios than frame machines, thus lower exhaust temperatures. They are lightweight and

Figure 3.1   A Mitsubishi-Hitachi M701J on the factory floor. *Source*: Reproduced by permission of Mitsubishi Hitachi Power Systems, Americas, Inc.

can be started very quickly, in as little as 10 min from cold conditions to full output, making them ideal for peaking applications where they can be used to take advantage of price spikes caused by power shortages. The largest aero-derivative unit is about 100 MW, but most are less than 50 MW. Due to their rapid starting capability, aero-derivative units are often used to supply backup power for renewable energy sources such as wind generators. They are often found as mechanical drives, refrigeration compressors, or in marine applications outside the power-generation industry.

Frame machines are large land-mounted units. Among the largest include the Siemens SGT5-8000H rated at 400 MW, and the Mitsubishi-Hitachi M701J rated at 470 MW shown in Figure 3.1. The M701J has a 15-stage compressor and four stage turbine section shown on the right in Figure 3.1. Frame machines generally have lower pressure ratios than aero-derivatives and consume a large volume of air. These characteristics yield high exhaust energy, making them ideal for combined cycle applications that include a steam turbine generator and are often used in CHP facilities providing steam to a host refinery or chemical plant.

Steady development of frame gas turbine technology has led to general classes of machines based mostly on the firing temperature, which is usually a trade secret. F-class machines, first introduced in the late 1980s, were a significant advancement over the E-class machines. The F-class fleet, nominally about 170 to 230 MW, had good commercial success in 50 Hz and 60 Hz markets. G-class machines, with a higher flame temperature, had only moderate commercial success and preceded the H-class introduced ca. 2007. Mitsubishi-Hitachi introduced the J-class frame gas turbine with a turbine inlet temperature of 1600 °C (2912 °F) in 2011, rated at 470 MW at 50 Hz.

In power generation, the maximum size of a gas turbine is determined in part by blade length, which limits the mass flow of air through the machine. Materials strong enough to withstand the high gas temperatures and stresses at full load limit blade length. Since a gas turbine's output is proportional to its inlet mass flow, those operating at 50 Hz have a higher output than the 60 Hz counterpart by the inverse square of the rotating speeds, or about 1.44 times the 60 Hz output.

3.1.2   Gas Compressor Fouling and Cleaning

Frame gas turbines ingest a huge quantity of air. A GE 7F.05® gas turbine (~223 MW) might consume as much as 508 kg/s of air, almost five times the volume inside Madison Square Garden every hour. Even filtered air can carry enough particulate matter to coat the surface of compressor blades and vanes and show noticeable gas turbine degradation within a few days. Compression slows the air speed across the blades, so the boundary layer and therefore the efficiency of compression is influenced significantly by deposits on the blades. A larger boundary layer and lower compression efficiency, requires more compression power from the turbine per kg of air consumed, and the turbine receives less gas flow from the combustors to produce its output.

Since the output of the gas turbine is proportional to the quantity of air, even small amounts of deposits on the compressor blades have a noticeable impact on the machine's output. Therefore, in addition to filtration, there are several traditional strategies employed to reduce deposit formation on the gas turbine compressor, including online and offline compressor washes.

Online washing involves spaying demineralized water directly into the compressor inlet while the machine is in operation. The effectiveness of online washing is often questioned as it cannot reach much beyond the first rotating blade row and all the water is evaporated due to the heat of compression by the fourth row of a compressor that may contain as many as 19 stages. Solids removed from the inlet guide vanes (IGVs) and the first rotating row often end up as deposits on the later stages of the compressor where they oxidize at temperatures approaching 370 °C, forming hard coatings on the blades. Additionally, water droplets entering the machine strike the leading edge of the rotating blades and quickly remove base metal leaving behind pits and sharp edges that tend to collect foreign matter, reduce the compressor efficiency, and can increased stress concentrations in areas already subject to high stresses.

Offline (crank/soak) washes are more effective than online washes and do not damage the leading edge of the first row. However, the machine must be removed from service, cooled, filled with water and detergent several times, dried, and returned to service. The entire process can be completed in a little over a day but, for planning purposes, 2 to 3 days are often reserved for offline washes. Depending on the material deposited on the blades, offline washes can become less and less effective due to hardening of the material on the later stages of the compressor.

The practices of online and offline washing are often requirements of the manufacturer or the major maintenance services provider. As manufacturers are the sole suppliers of parts for advanced technology gas turbines, they are also, very often, long-term services providers for gas turbine major maintenance. The maintenance contract can carry requirements to conduct daily online washes and periodic offline washes in order to maintain the service warranties. Because the manufacturer is the only parts provider for advanced technology models, the owner may not have much leverage in negotiating these types of requirements.

3.1.3   Exercise 1

Read the following articles: Kurz *et al.* (2011) and Stubbs (2009). Select the appropriate response for the following:

1. Particles less than ___ tend to stick to gas turbine blades.
    (a) 1μ
    (b) 5μ
    (c) 10μ
    (d) 12μ
2. Liquids remain on a compressor blade when the impact velocity is low.
    (a) True.
    (b) False.
3. Susceptibility to fouling depends mostly on:
    (a) Ambient conditions.
    (b) Engine size.
    (c) Engine compression ratio.
    (d) Filtration efficiency.
    (e) a. and b. above.
    (f) a. and d. above.
4. Genentech's condition based maintenance program reduced operating costs by:
    (a) Early replacement of freezer compressors.
    (b) Timely replacement of failing steam traps.
    (c) Resolving maintenance on freezer without replacement.
    (d) Avoiding losses of research samples.
    (e) b. c. and d. above.

3.1.4   Inlet Filtration

Typical components of gas-turbine inlet filtration systems in power generation applications include: weather hoods, moisture separation, and filter elements. There may be several stages of filtration with increasing efficiency for smaller particle sizes. Figure 3.2 is an elementary sketch showing the location of weather hoods, moisture removal, and various filter elements.

   Filter systems are classified as either static or self-cleaning types. Static filter systems contain elements that remain in service until they are dirty and ready for replacement. Self-cleaning type systems generally hold canister-type elements. The elements are designed to withstand a burst of air in the reverse direction to remove dirt and debris caught by the elements. Pulsing of the pneumatic bursts may be scheduled or dependent on a measured pressure drop across the filter elements.

   By improving the inlet filtration efficiency, compressor fouling due to particulates in the air can often be completely eliminated. In the mid-1990s, Mitsubishi began using HEPA inlet filtration for their large-frame gas turbines. Owners of these machines rarely, if ever, wash their compressors due to the high filter efficiency of HEPA filters.

   Figure 3.3 shows filter efficiencies for various classifications of filters over a range or particle sizes from 0.1 μm to 10 μm. An important distinction between HEPA filters and

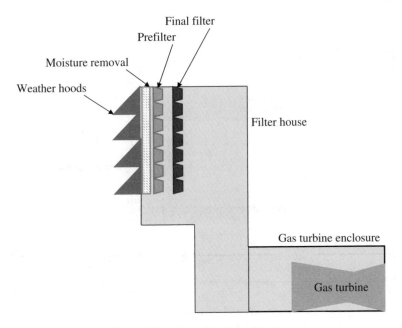

Figure 3.2 Gas turbine inlet filtration.

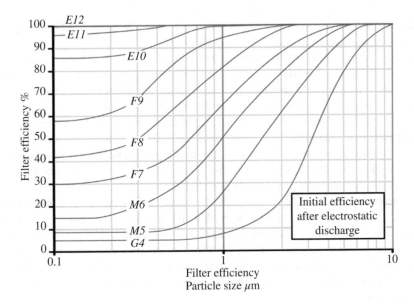

Figure 3.3 Filtration efficiency. *Source*: Reproduced by permission of Fram Industrial.

other classifications is that HEPA filters are rated based on their new and clean condition. Other filtration classifications are rated on a dirty or fouled condition where particulates have plugged many of the pathways through the filter media. Therefore, HEPA filters provide expected performance beginning upon initial installation. HEPA filters have rated

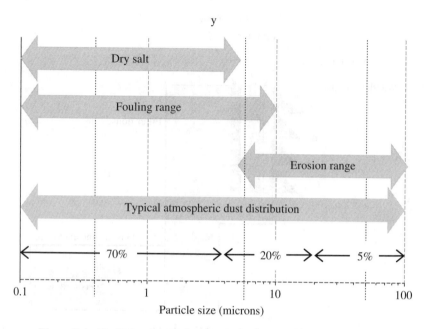

Figure 3.4   Typical atmospheric dust distribution. *Source*: SWRI/GMRC.

efficiency for particles as small as 0.01 μm, whereas F-class filters are rated in the 1 μm range, and G-class filters around 10 μm.

Figure 3.4 (Wilcox *et al.*, 2010) shows the typical particle size distribution for atmospheric air. As shown by Figure 3.3 and Figure 3.4, an F9 filter, a mainstay of gas turbine inlet filtration for years, would prevent erosion from taking place but about 30% of 0.03 μm particles attributed to compressor fouling would pass through the filter. In the case of a H11 HEPA filter, only about 3% of the 0.03 μm particles would pass through the filter when clean.

Finding the correct filter for a specific application can be complex. It requires knowledge of what is in the air, the size of the particles, seasonal changes in the air quality, and contamination present in the atmosphere that could degrade the filter media. Water-soluble contaminants trapped in filters – sea salt, for example – can liquefy and wick through the filter, entering the "clean" air stream. Contaminants such as salt, when dissolved in water, can initiate corrosion in the compressor, degrading the integrity of blades, and vanes, in addition to degrading gas turbine performance.

In some cases, a filter manufacturer may have a test skid that can be loaded with a preliminary selection of filter baskets and tested at the site to verify a preliminary selection of elements. Pressure drop through the mock-up filter can be monitored over a period and downstream air quality verified to assure the owner that the selected system will function as expected prior to purchase.

3.1.5 Gas Turbine Performance Measurement

The output of a gas turbine is roughly proportional to the mass flow through the machine. Since the volume flow through the machine is nearly constant, anything that affects the ambient air density – temperature, barometric pressure, inlet pressure drop, relative humidity – will have an effect on output and efficiency. Other factors that affect output and efficiency of a gas turbine include: shaft speed, fuel temperature, fuel composition (hydrogen to carbon ratio), the type of fuel – liquid or gas for a duel-fuel unit – exhaust back pressure, and others. A manufacturer's performance guarantee for a new unit will be valid only under specified ambient conditions, fuel quality, inlet and outlet pressure drops, running time on the machine, shaft speed, and perhaps others. Results of an acceptance test, conducted to verify a manufacturer's guarantee, must be corrected to the guarantee conditions specified using the manufacturer-provided correction curves. Likewise, regular performance monitoring mea-surements and results must be corrected to the same standard conditions for comparison. That is, the performance of a gas turbine operating in the summer at 33 °C cannot be directly compared to the same machine operating during winter conditions at 2 °C.

Usually, for performance monitoring and analysis:

- the inlet and outlet pressure losses are fairly constant;
- fuel temperature is controlled to a set value;
- pipeline-quality natural gas composition is nearly constant;
- shaft synchronous speed does not change appreciably; and
- the machine should operate at its specified firing temperature.

Therefore, the three main corrections to standard conditions can often be reduced to

- ambient temperature;
- barometric pressure; and
- relative humidity.

Of these, the primary correction is that for ambient temperature.

3.2   Case Study Details

The owner of the CHP plant under study is required by the gas-turbine maintenance contract to conduct daily water washes, and regular offline crack / soak washes on the three gas turbines at a facility it owns and operates. Offline washes are schedule after a gas turbine has lost bet-ween 5 and 10 MW due to compressor fouling. The manufacturer has recently recommended an upgrade to the online water-wash system that is advertised to reduce inlet blade erosion, improve spray effectiveness, and be more effective at removing deposits. Your firm is review-ing the operating practices, the manufacturer recommendations, and was asked to propose ways the owner would be able to maintain a higher output, efficiency, and steam reliability to the host facility. Since the frequency of offline water washes is uncertain, one method to improve the overall financial performance of the facility would be to optimize the frequency of washes and establish a procedure to ensure the optimum frequency is regularly followed.

There are two fairly straightforward methods to determine the optimum period between offline water washes: setting the derivative of the loss function equal to zero, or linear program-ming. In simple cases, either method is accurate. In more complex situations, with present values, discontinuous functions, and variable escalation rates, a full integration with a graphical solution may be more effective.

3.2.1 Derivative of the Cost Function

The owner has granted access to the CHP plant's online data historian and you have been able to capture a few months of hourly averages for gas-turbine temperatures and pressure for one of the units at the plant. This data begins at the end of April and extends to mid-July. From this data, you have culled the points when the gas turbine was not at 100% load and on the designated firing curve. The unit was removed from service on April 29 for an offline water wash and returned to regular full load service by May 3. The plant is located in the United States. Economic data provided by the owner includes:

- price of natural gas: \$4.50/MMBtu (\$15.35/MWh);
- discount rate: 11%;
- average market heat rate: 10.27 MMBtu/MWh (10.835 GJ/MWh);
- plant operating HHV efficiency: 53.56%;
- gas turbine and steam turbine output: 284 MW;
- gas turbine gross output: 179.1 MW;
- water wash outage: 48 hours;
- offline water wash fixed labor and material costs: \$20 000;
- reference ambient temperature: 50 °F (283.15 K);
- reference ambient pressure: 14.68 psia (101.24 kPa(a));
- average capacity factor: 92%.

Correction curves for the three identical gas turbines at the facility are shown in Figure 3.5, Figure 3.6, and Figure 3.7.

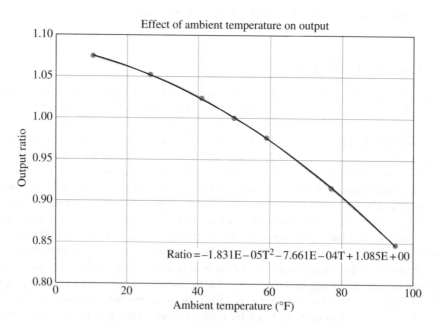

Figure 3.5 in the image shows a plot titled "Effect of ambient temperature on output" with the equation:

$$\text{Ratio} = -1.831\text{E}-05\text{T}^2 - 7.661\text{E}-04\text{T} + 1.085\text{E}+00$$

Figure 3.5   Ambient temperature correction.

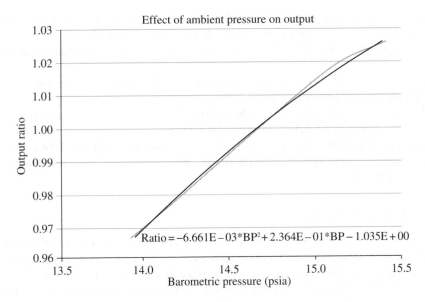

Figure 3.6   Ambient pressure correction.

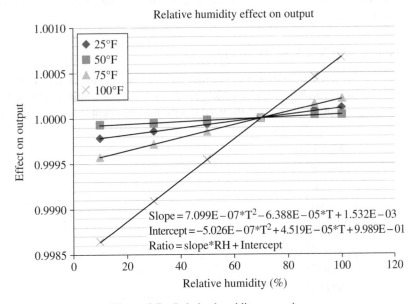

Figure 3.7   Relative humidity correction.

3.2.2   Exercise 2

1. Correct the measured gas-turbine output to standard temperature, barometric pressure and relative humidity.
2. Plot the lost gas-turbine output over the monitored period.

3. Determine an approximate rate of change for the gas turbine output due to compressor fouling.
4. Calculate the lost production costs for a 48-hour offline water wash.
5. Write a general equation for the losses associated with compressor fouling, and offline water washing.
6. Find the general expression for the optimum period between offline water washes during a year. Calculated the optimum period for the case study.
7. What are the annual saving for scheduling the offline water washes at the optimum cycle period rather than at a 5 MW loss on the gas turbine?
8. Discuss the variability of the test data. What factors may have influenced the scatter in the data and what techniques would you suggest to overcome uncertainties caused by the sample distribution?

3.2.3   Linear Programming

A linear programming model for this case is shown in Figure 3.8. In this method, it is easy to establish that the optimum water wash frequency occurs when the cost of lost generation due to fouling equals the cost to conduct the offline wash (lost production plus fixed labor costs.)

From either method, it is evident that the water-wash frequency is only a function of the rate of change in output due to compressor fouling. The greater the rate of change, the more frequent washes should occur at a constant price of fuel. The challenge, as seen above, was to determine the rate of change given the real nature of the fouling and its variable impact on the output of the machine.

3.2.4   New Methods – New Thinking

Instead of optimizing an existing situation, you may consider another alternative. Optimization assumes that the basic conditions under which a system functions continue. It is not wrong to maintain this assumption. It is, after all, what the gas-turbine manufacturer assumed given

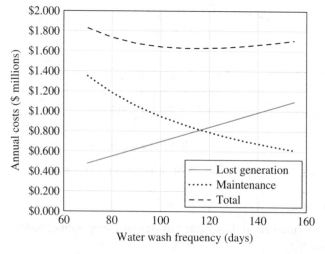

Figure 3.8   Linear programming solution.

the recommendation to purchase an upgraded water wash system. However, it is not the only alternative.

Another alternative is to assume the constraints of the existing system can be changed. That compressor fouling is not necessary for the future. That is, a given system of parameters does not have to remain the same, regardless of how long the process has continued. Working to implement a project that changes the given situation can be a better alternative than simply optimizing the given system. As mentioned above, a higher grade filtration system could completely eliminate fouling and the resultant lost generation, and the need for online and offline washes. The reliability of the facility can be improved as well as its long-term performance.

As stated earlier, the corporation has followed accepted good industry practices for operations and the maintenance of its facilities. However, good industry practices developed over time based on technologies present in the past. Rethinking the constraints provides opportunities to improve margins and opens new avenues for corporations to grow their business.

3.2.5   Exercise 3: Gas Turbine Inlet Filtration Upgrade

Analysis of competitive bids from two qualified supplies yielded a lowest evaluated cost of a new filter system of $5 000 000, including removal of the old system, installation, and testing of the new system.

- Determine the net present value of a capital improvement assuming that compressor fouling can be reduced to 10% of the present rate. Use the following inputs:
  - HEPA filter element life is the same as an F9 filter element: 3 years.
  - Differential filter cost: $100 000 compared to F9 filter elements.
  - Filter price escalation: 2.5% per year.
  - Natural gas and power price escalation: 0.75% per year.
  - Corporate tax rate: 28%.
  - Discount rate for evaluation: 11%.
  - Project life: 20 years.
  - The filter supplier will install a three-stage filtration system with a new filter house and connecting duct work during an upcoming scheduled outage. The regularly scheduled outage will need to be extended 18 days to allow for construction.
  - Under what conditions would you recommend a filter upgrade?

3.2.6   Presenting Results

A presentation of the results of this case-study analysis can be compressed into a few important points. A breakdown structure of the results can help organize the presentation and identify where the focus should be. Another method might be to create a storyboard of the presentation using the various elements that need to be presented. Writing down the elements on separate pieces of paper and using a magnetic board, you can rearrange the points, eliminate, modify or add points easily until a coherent presentation method is apparent.

There are two efficiency improvement alternatives to consider: one that can be implemented immediately, and the other an option for the near term. Figure 3.9 shows one possible

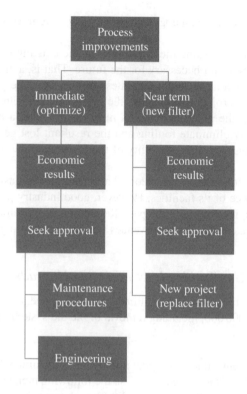

Figure 3.9   Example topic breakdown structure.

breakdown structure for the two options showing points to discuss and a suggested order. Once the topics are clearly presented, your company will have the opportunity to propose working with the client to complete one or both of the alternatives. Having a meeting with the client to discuss the findings, as opposed to a written report only, would create a means to discuss your proposal without interference from competing companies.

3.3   Case Study Results/Closure

The IPP in the case study justified replacement of the original, single-stage F9 filter system on the basis of gas-turbine reliability. The new three-stage filter system, which included a HEPA last-stage filter, required replacement of the existing filter house, and modifications to the inlet ducting. Prior to purchase, a test skid with the selected filter elements was mounted adjacent to the in-service filter, and operated at conditions simulating future operations for 6 months. Through the test period, the performance of the test skid verified the proposed design and showed no loss of filtration efficiency, and negligible increase in pressure loss over a wide range of ambient conditions.

Negotiations with the gas-turbine major maintenance provider allowed the owner to discontinue online and offline water washes pending the inlet filter upgrade to HEPA without affecting the warranty provisions, which was a major advantage of the IPP.

The filter manufacturer used computational fluid dynamic calculations to design the new filter house and inlet ducting to the gas turbine. The new, larger filter house provided smoother transitions in duct size and flow direction, thus limiting filtration pressure drop. Together with a redesigned inlet duct to the gas turbine, the pressure drop through the inlet system and duct work did not increase from the original design.

During the course of the first year of operation with the upgraded filters, online and offline compressor washing were halted. Gas-turbine compressor performance was carefully monitored showing that there was no detectable loss due to compressor fouling.

After 1 year of operation with the new filter system, and without compressor washing, the gas turbine manufacturer conducted an inspection of the compressor inlet. Inspections were to include replicate testing of the rotating blade leading edges to measure pitting damage, and sampling of blade deposits. The technician observed that the blades were as clean and in the same condition as the day they left the factory floor with no signs of deposits and no leading edge degradation. He immediately packed up his equipment and left the site. The owner declined the manufacturer's proposal to upgrade the online water wash system and, instead, disconnected the existing system.

While justification for the filtration system was based on reliability improvement, the complete absence of compressor fouling resulted in a positive net present value and a payback period of less than 5 years for the new filter system.

The company quickly upgraded the other two units at the site and leveraged the experience on four smaller units at another site. The four-unit facility did not require a new filter house and the filter supplier guaranteed performance with a single stage HEPA filter. The resulting payback period was just a few months.

During the first year of monitoring at the initial CHP facility, data scatter on two units essentially vanished while the third unit showed persistent data scatter. The scatter on the third unit was traced to the prevailing wind direction, which resulted in the gas turbine ingesting some of the cooling tower plume. When the third unit was constructed, it was placed in a nonoptimal location downwind from the cooling tower due to the land available. On occasion, unmixed and highly stratified air including the cooling tower plume would enter the unit making a determination of the temperature and relative humidity entering the gas turbine impossible. The unit performance was as variable as the wind. The units with the HEPA filters lacked deposits that were partly seasonal and partly influenced by online washes. Therefore the variability in the early readings was partly due to the blade deposits.

The filter media selected was hydrophobic to prevent water-soluble compounds from wicking through, and resistant to acids in the environment present from the combustion of fossil fuels. The hydrophobic media prevented sea salt and salts from nearby cooling towers from entering the clean air stream and protected the blades from corrosion. The media, continued to perform as guaranteed by the manufacturer throughout the warranty period and reorders with the same manufacturer were assured. Gas-turbine reliability improved in addition to a greater sales volume.

From an engineering perspective, seven projects were successfully implemented. The engineering firm was able to implement the modifications and registered repeat business with the owner. The owner, filter manufacturer, and engineering firm profited from the new business, and the engineer received a substantial year-end bonus.

3.4 Symbols and Abbreviations

b: intercept of the compressor fouling loss curve

m: slope of the compressor fouling loss curve

μm: micron

PH: period hours

SS: spark spread ($/MWh): gross margin from power sales = market price less the cost of fuel per MWh to produce the power (heat rate * fuel price)

WWD: water wash duration (h)

t: time between offline water washes (days)

t*:   Optimum period between offline washes (days)

LPO: Lost Profit Opportunity

MDC: Maximum Dependable Capability

FC: Fixed Costs.

3.5 Answer Key

Exercise 1

1. C
2. B
3. F
4. E

Exercise 2

1. A portion of the corrections spreadsheet calculation is shown below. Equations shown on the plots of correction curves are used to calculate the correction ratios. Corrected load = measured load divided by the ratios for ambient temperature, ambient pressure and relative humidity.

Elapsed Time	GTG Output	Inlet Temp	Rel. Humidity	Amb Press	Amb Temp	Amp Press	Rel. Humidity Slope	Rel. Humidity Intercept	Relative Humidity	Corrected load	Loss
(day)	(MW)	(F)	(%)	(Psia)	(Ratio)	(Ratio)			Ratio	(MW)	(MW)
0.0000	177.6	46.3	93.2	14.83	1.0103	1.0060	0.0001	0.9999	1.0000	174.7	4.4
0.0417	177.1	46.3	94.3	14.83	1.0103	1.0060	0.0001	0.9999	1.0000	174.2	4.8
0.0833	176.3	48.4	94.1	14.83	1.0050	1.0058	0.0001	0.9999	1.0000	174.4	4.6
0.1250	175.5	50.5	90.4	14.82	0.9997	1.0056	0.0001	0.9999	1.0000	174.6	4.5
0.1667	174.5	52.6	82.1	14.82	0.9941	1.0054	0.0001	0.9999	1.0000	174.6	4.4
0.2083	173.1	54.7	76.3	14.81	0.9883	1.0051	0.0002	0.9999	1.0000	174.3	4.7
0.2500	172.2	56.8	77.6	14.81	0.9824	1.0049	0.0002	0.9998	1.0000	174.4	4.6
0.2917	171.7	56.8	73.7	14.80	0.9824	1.0047	0.0002	0.9998	1.0000	174.0	5.0

2. Rate of change of load ~0.1 MW/day. See Figure 3.10.
3. Lost Profit Opportunity (LPO) for a 48-hour water-wash outage:

$$LPO = SS * MDC * WWD + FC = \$259,000$$

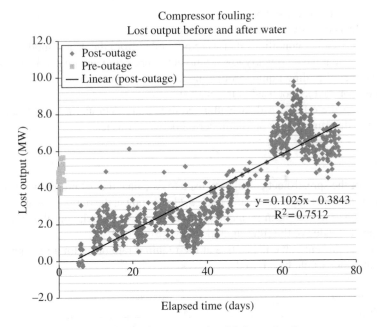

Figure 3.10   Answer: section 3.2.2, number 2.

where:

SS: spark spread or gross profit margin = market value of power – operating heat rate (MMBtu/MWh) * gas price ($/MMBtu);

MDC: maximum dependable capability = 284 MW;

WWD: water wash duration = 48 hours;

FC: fixed costs of water wash = $20 000.

4. Integrate the approximate loss curve and add the LPO and sum for one year:

$$\text{Loss} = \left\{ \left[1/2m(t - WWD/24) + b(t - WWD/24) \right] SS * CF + MDC * SS * WWD + FC \right\} 365/t$$

where:

t = days;

m = slope of the fouling loss curve (MW/day);

b = intercept of the fouling loss curve (MW).

5. The optimum interval between water washes:

$$t^* = \left\{ (WWD/24)^2 - 2 \left(b * WWD/24 - MDC * WWD/CF/24 - FC/CF/SS/24 \right) /m \right\}^{0.5}$$

t^* (optimum wash cycle) = 114 days

6. Savings at optimum rather than at 5 MW:
   (a) Losses at the calculated optimum: $1.63 million per year.
   (b) Losses at 5 MW loss – 51 days: $2.20 million per year.
   (c) Savings = $0.58 million per year.

7. Data scatter could be caused by:
   (a) The effects of daily online water washing.
   (b) Variable weather conditions that were not culled from the data – sudden showers, for example, that would quickly lower ambient temperature and raise relative humidity.
   (c) The effect of pulling liquid through the filter during rain showers.
   (d) Correction factors not used in the analysis.
   (e) Reducing the effects of data scatter requires more data and a longer sampling time prior to determining the optimum period between washes.

Exercise 3

1. A section of the spreadsheet calculation used to calculate the net present value of the filter installation and savings is shown in Figure 3.11. As shown, the optimum water-wash period for the HEPA filter has been recalculated to almost a year. Including the outage duration the net present value of the filter upgrade is over $600 000.
2. Since the net present value is positive at the required rate of return a recommendation to replace the filter would be advised.

HEPA differential cost	$ 100,000			
Filter guarantee period	3 years			
Escalation	2.50%			
Tax rate	28.00%			
Discount rate	11.00%			
Filter replacement option				
Year	**1**	**2**	**3**	**4**
Natural gas escalation (%/yr)	0.75%	0.75%	0.75%	0.75%
Power price escalation (%/yr)	0.75%	0.75%	0.75%	0.75%
Natural gas price ($/MWh)	$4.50	$4.53	$4.57	$4.60
Power price ($/MWh)	$46.22	$46.56	$46.91	$47.26
Production price ($/MWh)	$28.67	$28.88	$29.10	$29.32
LPC ($MWh)	**$17.55**	**$17.68**	**$17.81**	**$17.95**
Assumed rate of change for HEPA (MW/day)	0.0103			
Optimum water wash frequency (days)	361.3			
Annual losses current filter ($'000)	$1,627	$1,639	$1,651	$1,663
Projected annual losses HEPA ($'000)	$521	$525	$528	$532
Filter differential cost ($'000)	$ 100			$ 108
Pre-tax savings ($000)	$ 1,007	$ 1,115	$ 1,123	$ 1,023
Straight line depreciation of new filter ('000)	$250	$250	$250	$250
Taxes ($000)	$212	$242	$244	$216
Annual savings after tax ($'000)	**$757**	**$865**	**$873**	**$773**
Filter capital cost ($'000)	$5,000			
Outage extension cost ($'000)	$1,716			
Total filter cost ('000)	**$6,716**			
NPV@ 11% ($'000)	**$173.06**			

Figure 3.11   Net present value spreadsheet excerpt.

References

Kurz, R., Brun, K., and Mokhatab, S. (2011) Gas turbine compressor blade fouling mechanisms. *Pipeline and Gas Journal* **238**(9), http://www.pgjonline.com/gas-turbine-compressor-blade-fouling-mechanisms?page=show (accessed January 20, 2016).

Stubbs, C. (2009) *Implementing Performance Based Maintenance, Saving Energy and Improving Uptime*. Genentech, http://www.cypressenvirosystems.com/files/pdf/Genentech_Article_Final_Rev.pdf (accessed January 20, 2016).

Wilcox, M., Baldwin, R., Garcia-Hernandez, A., and Brun, K. (2010) *Guideline for Gas Turbine Inlet Air Filtration Systems*, Gas Turbine Research Council, Southwest Research Institute, http://www.gmrc.org/documents/GUIDE LINEFORGASTURBINEINLETAIRFILTRATIONSYSTEMS.pdf (accessed January 20, 2016).

Case 4

Flow Instrument Degradation, Use and Placement

The owner of a nuclear power plant has a responsibility to operate the facility in strict accordance with permits granted by local, state, and federal authorities including provisions set forth in the Final Safety Analysis Report (FSAR). One such provision is to operate the nuclear reactor at no more than 100% of its licensed thermal output power (MW_t). One method of determining the reactor core power level is the continuous measurement of the thermal input to the steam power cycle – steam calorimetry.

The full-time operation of a nuclear power plant at 100% reactor, or core, power is often the desire of its owner due to the low marginal cost of production. High-accuracy measurements of core power ensure that the owner operates as close to the full power level as possible, and reduces the size of emergency systems designed to cool the reactor under the most extreme conditions. Steam-side calorimetry is very often one of the means of measuring the reactor's core power and may be compared with others, depending on its established accuracy. The primary contributors to the uncertainty in steam side calorimetry are the measurements of feedwater flow and temperature.

A regulated utility owns and operates a nuclear power plant consisting of two pressurize water-reactor (PWR) nuclear units, each producing 780 MW of electric power (MW_e). Fifteen years after initial operations, the units are operating reliably for long periods at full load. The exception to the good performance is that the electric output is about 2% below what is expected. Records show that the output fell slowly over the past 15 years but it seems to have leveled off at a loss of about 2%. After a major refueling outage of one of the two units, which included an overhaul of the steam turbine generator, and inspection of the steam generating unit (SGU,) the output loss of 2% was confirmed by a performance test. Circumstantial evidence of a systematic loss in output was provided by the rate of fuel depletion during the previous fuel cycle.

Case Studies in Mechanical Engineering: Decision Making, Thermodynamics, Fluid Mechanics and Heat Transfer, First Edition. Stuart Sabol.
© 2016 John Wiley & Sons, Ltd. Published 2016 by John Wiley & Sons, Ltd.
Companion website: www.wiley.com/go/sabol/mechanical

The plant is an important source of revenue for the utility and its continued superior performance is important for investors who purchased bonds for its construction. The state's public utility commission (PUC) is keenly aware of the facility and its performance. The PUC demands that the utility operate it in a safe, reliable, and efficient manner for the overall benefit of the electric rate payers. Therefore, even the slight loss of output is of paramount importance, which led the senior vice-president of operations to ask the engineering department to investigate the problem and provide a solution.

4.1 Background

4.1.1 Nuclear Steam Power Cycles

There are two basic nuclear steam cycle designs: the boiling water reactor (BWR) shown schematically in Figure 4.1 and the pressurized water reactor (PWR), Figure 4.2. Boiling water reactors generate steam for power generation directly in the nuclear reactor. The boiling water has a high heat-transfer coefficient and therefore provides efficient cooling for the reactor. This efficient cooling lowers the cost of the construction of the reactor core but the steam system carries radioactive material. Therefore, power generating equipment must be shielded for normal operations and maintenance, which results in additional costs.

In a PWR system, a primary cooling loop of pressurized water cools the reactor without boiling. The primary cooling loop is cooled against the secondary loop in the steam generating unit (SGU), where water is boiled to provide steam for power generation.

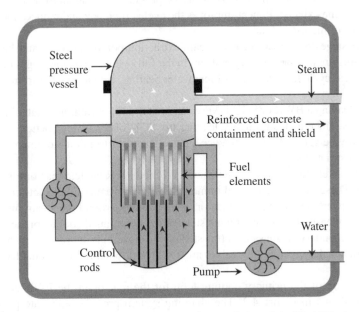

Figure 4.1   BWR configuration. *Source*: Reproduced by permission of the World Nuclear Association.

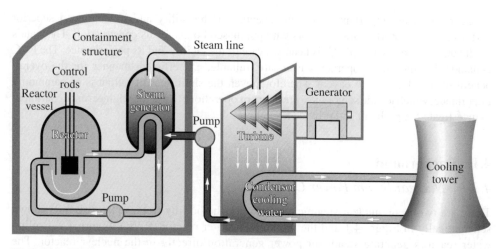

Figure 4.2   PWR configuration. *Source*: Reproduced by permission of the World Nuclear Association.

The lower overall heat-transfer coefficient provided by the pressurized water loop increases the cost of the nuclear reactor system but provides separation between the primary reactor coolant and the secondary power cycle. This separation lowers the cost of the power-generating system as compared to the BWR design making the two systems more cost competitive.

The BWR and PWR power cycles share common characteristics. The main steam that is generated in both cycles is saturated steam without superheat and mostly below about 6.9 MPa (1000 psi). Main steam is often used to reheat low-pressure (LP) steam from the high-pressure (HP) steam-turbine exhaust. Reheat improves the cycle efficiency and reduces the moisture in the final stages of the LP turbine.

In the later stages of the LP turbine, steam extraction points for feedwater heater remove condensed moisture that provide dryer steam to the following expansion stages of the turbine. A typical nuclear cycle has six stages of feedwater heating, with as many as four having moisture removal.

The steam leaving a BWR or SGU is very nearly at the vapor saturated state. Manufacturers can usually guarantee a steam quality leaving the SGU of 99.75% or better. The guarantee would normally be verified during the acceptance test prior to initial commercial operations of a newly constructed facility.

Of the many safety features and systems of BWR and PWR nuclear reactors are main steam dump valves. These valves are located immediately downstream of the reactor or SGU. Their function is to divert steam from the reactor or SGU outlet around the steam turbine directly to the steam condenser. Diversion of main steam allows continued cooling of the reactor in the event that the secondary steam system (the steam power cycle) is out of service. Although they are an important safety feature, the valves are a source of potential losses in electrical output if they leak.

Another feature of the primary cooling loop for the reactor in the PWR is the charging system – not shown in Figure 4.2. The charging system provides adequate pressure in the primary loop to prevent boiling of the water in the primary loop.

4.1.2   Core Power-Level Measurement

Calculations of the secondary side thermal input (steam side calorimetry) of a PWR rely on a primary flow measurement – most often feedwater flow to the steam generator. Also required are:

- the change in enthalpy through the SGU;
- a measurement of steam generator blowdown flow; and
- Inputs from the reactor coolant system and pressurizer heater power.

Traditional methods of measuring the feedwater flow relied on an ASME PCT6 throat tap flow nozzle or custom flow tube located in a calibrated flow section. The flow section would include the upstream and downstream piping, a flow straightener, and the differential pressure flow element itself. The downstream pressure measurement is located within the nozzle, at the throat or exit of the nozzle. The upstream pipe section would be bored to an exact diameter to remove any uncertainty due to manufacturing tolerances and to provide a smooth surface upstream of the flow element. The element and flow section would be calibrated against a standard in a hydraulics laboratory prior to shipment for installation. The calibration would have a guaranteed accuracy traceable to a national standard. Similar ASME PTC 6 flow sections are employed in fossil as well as combined cycle power generating units.

The steam-side calorimetric calculations for the BWR or PWR would be programmed into the plant control system prior to commissioning. Constants used in the calculation, including flow nozzle characteristics, reactor coolant pump heat input, and steam quality, would be verified by acceptance tests in the field or prior to shipment from the factory. An uncertainty analysis of the calculation would be completed by the engineer of record for the facility design, and included in the FSAR. The Nuclear Regulatory Commission would require an independent review of the FSAR prior to their approval, and issuance of the license for initial operation of the facility.

During normal operation, technicians would regularly calibrate temperature, pressure, and differential pressure transmitters required by the calorimetry to assure an accurate determination of the reactor core power. Operators would rely on the computer-generated results as accurate and dependable for the continuous, safe operation of the plant.

4.1.3   Differential Pressure Flow Measurement Devices

There are a variety of differential pressure flow measurement devices used to measure the flow rate of liquids and gases. For accuracies sufficient for the nuclear core power calorimetry, a calibrated ASME PTC 6 flow nozzle such as the one pictured in Figure 4.3, or custom flow tube would typically be used. In recent years, other methods, including sonic measurement using sensors mounted externally on the pipe, have become available and approved by ASME. ASME Performance Test Codes (PTC) 19.5 and PTC 6 together with ASME MFC-3M "Measurement of Fluid Flow in Pipes Using Orifice, Nozzle, and Venturi" are excellent references for the manufacture, installation, and use of all types of flow-measurement devices.

It is not the intent of this text to duplicate the information contained in the ASME documents. Independent reading and research into these and other industry standards is encouraged. There

Figure 4.3 ASME flow nozzle. *Source*: Reproduced by permission of Triad Measurement and Equipment.

are a few practices, and equations from the references, which are pertinent to this case study. In 1984, ASME PTC 6 allowed use of a primary flow element in the final feedwater, downstream of a deaerating feedwater heater, for acceptance testing as long as an inspection port was included in the flow section. To maintain the highest accuracy, the nozzle was to be inspected prior to, and after the performance test, to ensure the nozzle was not damaged, and did not require cleaning. Any deposits on the flow nozzle prior to the test were to be cleaned with a high-pressure water-jet device. Inspections showing an iron oxide deposit should ensure the thickness was less than specified by PTC 6 and that the surface of the nozzle remained smooth throughout performance testing. A rough surface would decrease the nozzle discharge coefficient, and result in an indication of flow that was higher than the true flow through the nozzle.

For continuous real-time measurement of feedwater flow required for steam calorimetry in a nuclear power generating facility, regular inspections of the feedwater nozzle are impractical. In this case study, the inspection ports were not installed.

Equation (4.1) shows the general equation for flow through a differential flow meter (from ASME 19.5):

$$\dot{m} = n\pi \frac{d^2}{4} CY \sqrt{\frac{2\rho\Delta P g_c}{\left(1-\beta^4\right)}} \tag{4.1}$$

Values and units for the parameters in equation (4.1) are given in Table 4.1.

For the measurement of two-phase flow, as in the case of the flow from an SGU in a nuclear plant, ASME PTC 6 includes a modified flow equation (equation 4.2):

$$\dot{m} = \frac{n\pi \dfrac{d^2}{4} CY}{\sqrt{\left(1-\beta^4\right)}} \cdot \sqrt{\frac{\Delta P}{x^{1.5} \cdot \left(\upsilon_g - \upsilon_f\right) + \upsilon_f}} \tag{4.2}$$

Table 4.1 Flow equation units of measure.

Symbol	SI	Customary
$\dot{m}$	kg/s	lb_m/h
n	1 (kg/(m s² Pa)	300 (ft²/s²) [(in² s²)/(ft² h²)]$^{0.5}$
D	m	in
C	Dimensionless	Dimensionless
ρ	kg/m³	lb_m/ft^3
P	Pa	psi
g_c: proportionality constant	1 dimensionless	32.174 [(lb$_m$ ft)/(lb$_f$ s²)]

Source: From ASME 19.5.

When measuring gas flow with throat tap nozzles, and venture tubes, the adiabatic expansion factor (Y) is found from equation (4.3) per ASME MFC-3M. For liquids, Y is equal to one.

$$Y = \left\{ \left(\frac{\kappa \tau^{2/\kappa}}{\kappa - 1} \right) \left(\frac{1 - \beta^4}{1 - \beta^4 \tau^{2/\kappa}} \right) \left[\frac{1 - \tau^{(\kappa-1)/\kappa}}{1 - \tau} \right] \right\}^{0.5} \tag{4.3}$$

The uncertainty of the Y, also from ASME MFC-3M is shown by equation (4.4).

$$\text{relative uncertainty of } Y\left(\%\right) = \pm\left(4 + 100\beta^8\right)\frac{\Delta P}{P} \tag{4.4}$$

Restrictions for the use of equations (4.3) and (4.4) are set forth in ASME MFC-3M.

The isentropic exponent, κ, in equation (4.3) for steam can be approximated by taking a finite differential from the equation (4.5). Some steam-table functions available for use in spreadsheets can provide the isentropic exponent given for specified thermodynamic states:

$$\kappa = \frac{\upsilon}{P} \cdot \left(\frac{\partial P}{\partial \upsilon} \right)_s \tag{4.5}$$

Between 4 MPa and 7.6 MPA and steam qualities (x) between 0.9 and 0.9994, κ can be approximated as a function of pressure in MPa with equations (4.6) and (4.7)

$$\kappa = a_2 \cdot P^2 + a_1 \cdot P + a_0 \tag{4.6}$$

where

$$a_0 = 0.1194 \cdot \ln\left(x\right) + 1.1533$$
$$a_1 \cdot 10^2 = 1.6133 \cdot \ln\left(x\right) - 1.114 \tag{4.7}$$
$$a_2 \cdot 10^4 = -7.753 \cdot \ln\left(x\right) - 2.3415$$

Pipe and flow element diameters are measured at ambient temperatures. These measurements must be corrected for operating temperatures. The lineal thermal expansion factor for

Table 4.2   Constants for linear thermal expansion factor for temperature in (°C).

Material	a	b	c	d
Austenitic stainless steel	16.224	0.0063076	−5.9575E-06	3.6098E-9
Carbon: carbon-moly steels	10.728	0.0081725	−1.6951E-6	−2.0374E-9

Source: ASME PTC 19.5.

various materials is given by the equation (4.8 with constants provided in Table 4.2 (ASME19.5).

$$\alpha \cdot 10^6 = a + bT + cT^2 + dT^3 \tag{4.8}$$

Length at the operating temperature is calculated with equation (4.9) from ASME PTC 19.5:

$$L_T = L_B \left\{ 1 + \alpha \left(T - B \right) \right\} \tag{4.9}$$

where:

L_T = length at operating temperature;
L_B = length at base temperature;
T = operating temperature (°C);
B = base temperature, generally 20 °C.

When a flow section containing a differential flow element is calibrated, the laboratory usually establishes the flow element discharge coefficient at a throat Reynolds number of about one million. The operating Reynolds number is most often many times higher than during calibration; therefore, the calibrated value for C must be adjusted to the operating conditions. For a calibrated ASME PTC 6 flow nozzle or custom flow tube, the constant C_x is calculated to satisfy equation (4.10) for the calibrated values of C. Each value of C from an individual calibration, determines a value for C_x at the corresponding Reynolds numbers. There may be several different calibration tests required for a single nozzle and each set of upstream and downstream pressure taps. The average value for C_x for a specific set of pressure taps is then used to find the value of the discharge coefficient, C, at any Reynolds number greater than one million using equation (4.10).

$$C = C_x - 0.185 R_d^{-0.2} \left(1 - 361,239 / R_d \right)^{0.8} \tag{4.10}$$

(ASME PTC 6)
where:

C: discharge coefficient at specific throat Reynolds numbers;
C_x: a constant.

Equation (4.10) results in an iterative calculation of flow. The iteration procedure begins with the calibrated value of C, yielding a calculated flow rate. At the Reynolds number determined at the flow rate, C is re-established, and a new flow calculated. The process repeats until the flow rate converges to a satisfactory tolerance. Convergence is generally very rapid.

The permanent pressure loss from an ASME flow nozzle can be approximated per ASME 19.5 according to equation (4.11). If the flow section includes a diffuser behind the nozzle, approximately 70% of the loss may be recovered. If a venture or custom flow tube with a diffuser is used, ASME standards recommend that between 5% and 20% of the measured differential pressure will be a permanent pressure loss.

$$\frac{pressure\ loss}{differential\ pressure} = 1 + 0.014\beta - 2.06\beta^2 + 1.18\beta^3 \qquad (4.11)$$

ASME MFC-3M suggests that the permanent pressure loss through a nozzle can be approximated by equation (4.12). Equations (4.11) and (4.12) yield similar results:

$$pressure\ loss = \frac{\sqrt{1-\beta^4(1-C^2)} - C\beta^2}{\sqrt{1-\beta^4(1-C^2)} + C\beta^2}\Delta P \qquad (4.12)$$

4.1.4 Two-Phase Piping Pressure Drop

The flow of two-phase flashing or condensing flows in pipes is not well understood. There are numerous formulations, theories, and approximations for calculating the pressure loss in two-phase systems, most of which are formulated for a specific set of fluids and conditions. The consistency among the formulations is that the pressure loss for a two-phase system is greater than for a comparable single-phase flow. For this case study, the two-phase pressure drops are provided from earlier calculations, and conservative values for the uncertainty of the pressure losses are applied to account for the unknowns associated with two-phase pressure loss calculations.

Data is provided in the details of the case study to allow a calculation of the two-phase friction and static pressure losses, if desired. The author suggests following Müller-Steinhagen and Heck for two-phase friction pressure loss and Rouhani and Axelsson (1970) equation I for C, which can be found in Woldesemayat and Ghjar (2006).

4.1.5 Uncertainty

Uncertainties that should be considered in the measurement of flow include:

- Nozzle and piping diameter. For calibrated flow sections, the upstream piping should be bored to a known diameter. The measurements should be accurate to ±0.00127 mm (±0.0005").
- Calibrated discharge coefficients: the typical laboratory accuracy of measuring the discharge coefficient is 0.25% of the measured value.
- Pressure measurement. Many transducers produce an analogue current signal (often 4 to 20 mAmp) proportional to the measurement. The total uncertainty of the measurement must account for the transducer, transmission of the signal, and conversion to a digital value that can be interpreted by a computer control system.

- Temperature measurement: standard accuracy J type thermocouples have a published accuracy of 0.77% of the reading. As with a pressure signal, there would be instrument-loop uncertainty in addition to the uncertainty contributed by the thermocouple.
- Adiabatic expansion factor: the uncertainty of the adiabatic expansion factor is given by Equation (4.4).
- Calculation of two-phase pressure loss. For this case study, the author recommends ±20% of the calculated pressure loss.
- The measurement of blowdown flow is assumed to have an accuracy of ±15%.
- After factory acceptance of the reactor coolant system components, the uncertainty of inputs to the steam calorimetry calculation is assumed to be ±5%.

4.2   Case Study Details

After a brief review of the background material and proposed theories for the loss of power, you have taken a trip to the site to collect operating data, drawings and documentation. Your findings are shown below:

- The plant has three steam generators (SGUs) per turbine. There are two 1800 RPM, four-pole steam-turbine generators, each rated at 780 MW. Both steam turbine generators have shown a similar loss in output.
- The main steam lines are custom-made 0.7112 m (28") inside diameter lines and there is approximately 152 equivalent meters (500 ft) of pipe from the SGU outlet to the steam flow meter section and another 18 m (60 ft) of straight pipe downstream of the meter to the pressure measurement used to measure SGU outlet pressure. The SGU outlet is at an elevation of 39.6m (130 ft) and the steam flow nozzle is in a horizontal section located at 9.1 m (30 ft) elevation.
- Steam quality from the steam generators was measured at 99.94% using radioactive tracers during the initial performance test prior to commercial operations.
- Average steam flow: 1320 kg/s (10 500 000 lb$_m$/h).
- Average feedwater flow 1350 kg/s (10 700 000 lb$_m$/h).
- Main steam dump line temperatures at or near ambient temperature.
- At the time of its installation, the plant computer / control system was state of the art for the control and monitoring of nuclear power plants with 26k of random access memory. There are limited engineering functions available, most of which are in assembler language. A Fortran compiler is installed on the system.
- You find the computer program to calculate reactor core power in customary units is as shown by equation (4.13):

$$Core\,Power = \sum_{i=1}^{3} \left[\frac{\dot{m}_{FWi}\left(h_{Si} - h_{FWi}\right) - \dot{m}_{BDi}h_{Si}}{3412.14 \cdot 1000} \right] - CHG - RCP \qquad (4.13)$$

where:
CHG = charging and letdown flow heat additions (constant 1.1 MWt);
RCP = reactor coolant pump heat addition (constant 35 MMBtu/h);
i = SGUs 1, 2, and 3.

- SGU outlet gage pressure = 5.92 MPa (858 psi).
- SGU outlet temperature = 549 K (529 °F).
- SGU blowdown flows are measured using flow orifice meters and are generally about 0.3% of feedwater flow.
- The steam pressure, and flow differential pressure transmitters have a stated accuracy of 0.055% of span. Pressure gages are calibrated from 0 to 6.89 MPag (1000 psig) and differential pressure transmitters between 0 and 125 kPa (500 "H$_2$O).
- Temperatures are measured with standard type-J thermocouples.
- Instrument loop accuracies have been shown to add 0.02% to the transmitter uncertainties.
- The design engineering firm for the facility recommended insulation losses of 1.5 MW$_t$ per reactor.
- The plant computer system collects 1 s samples of data and averages these values for 1 min – using the 1 min averages to calculate core power.
- Steam and feedwater flows are measured by calibrated ASME throat tap nozzles. The steam nozzles were installed with diffusers. All nozzles were made of 316 SS, installed in bored piping sections with inline flow straighteners – one nozzle in each line to the three steam generators per steam turbine. The nozzles are ASME PTC 6 instruments and the flow sections were calibrated at a hydraulics laboratory that certified their discharge coefficient to within ±0.25% at a Reynolds number near 1 000 000.
- Feedwater and steam piping are either carbon steel or a chromium / molybdenum steel.

To eliminate the possibility of systematic errors caused by simplifications in the plant computer system calculations, you collected the recorded data for the feedwater and steam flow measurements shown above. You also found the original calibration sheets for each nozzle showing the specifics of the constructions and calibration results. The data is summarized below for the three elements labeled 1, 2, and 3 for one of the two reactors at the site.

To substantiate a statistical difference between feedwater flow and steam flow, you will need to calculate the measurement uncertainty for each. Noting the number of parameters in the flow calculation, you have decided to use a Monte Carlo simulation to determine uncertainty. You have concluded that, if the uncertainty bands for steam and feedwater flow measurements do not overlap, instrument accuracy would show a consistent difference that could explain the lost power.

4.3   Exercises

Complete the following:

1. Review and evaluate equation (4.13). What changes would you recommend?
2. Write the core power calorimetric equation using steam flow rather than feedwater flow.
3. Using the values in Table 4.3, and the measurement uncertainties, determine if steam temperature is more accurately determined by the pressure or temperature measurements.
4. Using the values in Table 4.3, calculate the following for the three feedwater flow nozzles:
   (a) Operating beta ratio.
   (b) Nozzle discharge coefficient at the operating flow.
   (c) The flow rate through each nozzle.

Table 4.3 Flow measurements and instrumentation.

SGU	1	2	3
Feedwater Flow Nozzles			
Pipe bore (m/in)	0.42558/16.755	0.42553/16.753	0.42565/16.758
Nozzle diameter ratio	0.4886	0.4953	0.4925
Flow coefficient at Rd = 1 000 000	0.9949	0.9955	0.9952
Pressure (MPag/psig)	6.151/892.1	6.127/888.6	6.140/890.5
Temperature (K/°F)	499/438	500/440.1	498/437
Differential pressure (kPa/"H_2O)	98.20/394.2	93.27/374.43	94.70/380.2
Steam Flow Nozzles			
Pipe Bore (in)	0.71628/28.200	0.71628/28.200	0.71628/28.200
Nozzle diameter ratio	0.6464	0.6466	0.6464
Flow coefficient at Rd = 1 000 000	0.9970	0.9973	0.9943
Pressure (psig)	5.915/857.9	5.914/857.7	5.912/857.5
Temperature (°F)	549/527.8	551/531.3	549/5329.2
Differential pressure ("H2O)	92.15/369.9	92.27/370.4	91.97/369.2

5. Using the values in Table 4.3, calculate the following for the three steam-flow meters:
   (a) Operating beta ratio at the operating temperature.
   (b) Isentropic exponent (κ).
   (c) Adiabatic expansion factor.
   (d) Two-phase flow correction.
   (e) Operating nozzle discharge coefficient.
   (f) The flow rate through each nozzle.
6. Calculate reactor core power
   (a) Using feedwater flow.
   (b) Using steam flow.
7. From Figure 4.4, what changes should be made to the instrumentation to improve the accuracy of the calculation of core power?

4.3.1 Uncertainty

Uncertainty may be determined by application of the Kline and McClintock equation or through a Monte Carlo analysis. The Kline and McClintock method requires determinations of the partial derivatives of the core power with respect to each of the measured parameters, pressure, temperature, differential pressures, pipe diameter, and so forth. This can be an exhaustive exercise, especially in the case of the steam flow measurement.

The alternative is to perform a Monte Carlo simulation using a random-number generator to determine the extent of the uncertainty. For an explanation of a Monte Carlo simulation, see:

http://en.wikipedia.org/wiki/Monte_Carlo_method (accessed January 22, 2016)

For this case study, follow the steps outlined below to complete a Monte Carlo simulation of the core power calculation completed in 4.3 above by writing a computer program:

1. Create a table of maximum uncertainties of each measured parameter and calculated values where appropriate.

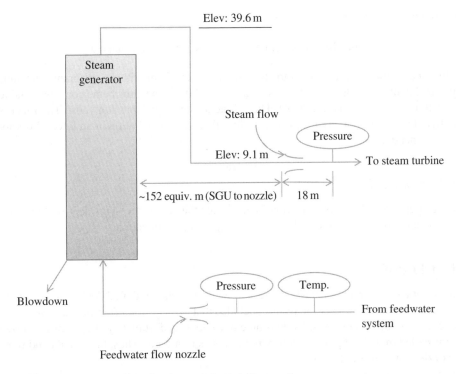

Elev: 39.6 m

Steam generator

Steam flow

Pressure

Elev: 9.1 m

To steam turbine

~152 equiv. m (SGU to nozzle) | 18 m

Pressure   Temp.

Blowdown

From feedwater system

Feedwater flow nozzle

Figure 4.4   SGU schematic.

2. Set up a random number for each parameter between the zero and one.
3. Calculate the uncertainty as a function of the random number. For a random number of zero the uncertainty would be the maximum negative uncertainty, at one the maximum positive uncertainty.
4. Add or multiply the uncertainties to the measured parameters depending on how they are defined.
5. Calculate the core power using the steps in section 4.3 above.
6. Repeat steps 2. through 6. until the average value of core power converges with the value calculated in section 4.3, about 2000 times.
7. Determine the standard deviation, maximum and minimum value of core power from the trials of step 6. using feedwater flow and steam flow. Select the maximum of three times the standard deviation or the absolute maximum and minimum as the range of uncertainty of the calculation of core power.

When using Excel to perform the Monte Carlo, insert an ActiveX command button in the spreadsheet using the Insert function on the Developer toolbar. Access the Visual Basic editing tool to write the subroutine initiated by the Active-x control outlined above. Random numbers require invoking the "Randomize" function prior to generating random numbers with the function "RND."

Cell values within worksheets may be accessed within the subroutine with the following statement:

$$x = \text{Worksheets}(\text{"sheet name"}).\text{Cells}(\text{row, col})$$

and written to cells with:

$$\text{Worksheets}(\text{"sheet name"}).\text{Cells}(\text{row}, \text{col}).\text{Value} = x$$

This calculation requires access to steam-table values from within a program. The author suggests downloading Excel functions available at http://www.mycheme.com/steam-tables-in-excel/(accessed January 22, 2016). These functions are available at no cost. The functions are offered as-is without any responsibility for errors. For installation in an Excel Workbook see the Appendix.

4.3.2   Conclusions

Write a short PowerPoint presentation intended for the senior vice-president outlining your findings, conclusions, recommendations and proposed plan of action.

4.4   Closure

Analysis of feedwater and steam-flow measurements together with fuel depletion, and nuclear steam cycle performance cycle analysis concluded that the feedwater flow nozzles had become fouled. Iron oxide was the most likely source of fouling. Unfortunately, the feedwater nozzles were installed in welded pipe sections without inspection ports. Thus, the actual condition of the nozzles was never determined.

The online calculation of reactor core power was rewritten to use steam flow rather than feedwater flow. Due to the limitation of the plant computer, many of the parameters for steam flow and blowdown flow were calculated at full load, and a single constant used with the measured differential pressure. The calculation was corrected with the addition of insulation losses. The FSAR was updated and submitted to the Nuclear Regulatory Commission (NRC) with a calculated uncertainty of the core power calculation of $\pm 0.6\%$. The NRC's approval was received shortly after submission.

At the earliest opportunity, pressure gages were added in the main steam lines from the separate SGUs. One was mounted immediately downstream of each SGU outlet and one upstream of the each steam flow nozzle per ASME guidelines. Once installed, the computer program was updated to reflect the new instruments.

At an electricity price prior to adjustments for transmission and distribution of $40/MWh, the change to steam flow has returned almost $300 million to the utility's customers since the change was made.

4.5   Symbols and Abbreviations

A: area, L^2
C: flow element discharge coefficient
d: flow element throat diameter, L
D: pipe diameter, L
Force: $M\,L/t^2$
g_c: proportionality constant

h: specific enthalpy per unit mass
L: length, distance
$\dot{m}$:   mass flow, M/t
M: mass
n: units conversion
P: pressure, F/A
R_d: Reynolds number at diameter "d"
s: specific entropy per unit mass
t: time
T: temperature
v: specific volume (L^3/M)
x: steam quality (−)
Y: adiabatic expansion factor
α: linear thermal expansion factor
β: ratio of flow element throat diameter to pipe diameter D
γ: ratio of specific heats
Δ: finite differential operator
κ: Isentropic exponent
ρ: fluid density, M/L^3
τ: pressure ratio (P_2/P_1)

Subscripts
1, 2, 3... Sequential states or events

BD: SGU blowdown
f: liquid phase
g: gas phase
i: sequence as in the first, second and third SGU
FW: feedwater to the SGU
S: Steam from the SGU

4.6   Answer Key

Section 4.3

1. Equation (4.13) contains two errors. The first is that the blowdown flow is multiplied by the main steam enthalpy and not the change in enthalpy from saturated liquid to the main steam enthalpy. Though the calculation was prepared by a competent manufacturer, reviewed by the engineer of record designing the facility, with a second review by an independent engineer qualifying the calculation for the Nuclear Regulatory Commission, the calculation passed and was used for approximately 15 years before the error was detected. The error decreased the value of core power by a small amount allowing the owner to operate the facility slightly above the licensed limit. The second error was the exclusion of insulation losses. Again this exclusion diminished the calculated value of reactor core power and allowed operation slightly above the licensed limit.

Table 4.4   Answer, section 4.3, number 4.

SGU	1	2	3
Feedwater Flow			
β	0.4891	0.4958	0.4930
C	0.9969	0.9975	0.9972
$\dot{m}$ (kg/s/kpph)	449.7/3569	450.8/3578	449.6/3568

Table 4.5   Answer, section 4.3, number 4.

SGU	1	2	3
Steam Flow			
β	0.6472	0.6474	0.6472
κ	1.077	1.077	1.077
Y	0.9864	0.9864	0.9864
Two-phase	1.0009	1.0009	1.0009
C	1.0001	1.0004	0.9974
$\dot{m}$ (kg/s/kpph)	441.1/3501	441.8/3507	439.3/3487

2. Using steam flow, reactor core power is calculated as follows in equation 4.14:

$$Core\,Power = \sum \left\{ \frac{\left[\dot{m}_{Si}\left(h_{Si} - h_{FWi}\right) + \dot{m}_{BDi}\left(h_{BDi} - h_{FWi}\right) \right]}{3412.14 * 1000} \right\} - CHG - RCP + Ins \quad (4.14)$$

Where: Ins = Insulation losses
3. SGU Steam temperature determination.
   (a) There are three possible uncertainties that contribute to the measurement of the SGU outlet pressure, including: the measurement of pressure downstream of the flow nozzle, the flow nozzle pressure loss and the piping pressure losses between the SGU outlet and the flow nozzle. The absolute maximum uncertainly of the SGU pressure would be ±22 kPa (±3.2 psia). Steam-saturated temperature calculated from pressure would have an uncertainty of ±0.2 K (±0.4 °F).
   (b) The thermocouple accuracy is ±2 K (±4 °F).
   (c) Therefore, using the pressure measurement to calculate the saturation temperature is more accurate than a direct measurement of temperature for this case.
4. Feedwater Flow see Table 4.4.
5. Steam Flows see Table 4.5
6. Core Power using feedwater flow = 2407.7. Core Power using steam flow = 2364.6.
7. Changes to instrumentation:
   (a) Add pressure measurement upstream of each steam flow nozzle.
   (b) Add pressure measurement at outlet of each SGU.

Section 4.3.1, Monte Carlo simulation.

8. The minimum core power calculated from feedwater flow is about 0.58% above the maximum core power from steam flow. It is therefore unlikely that measurement uncertainty has led to the difference in the two measurements.

   (a) Based on feedwater flow from 4000 trials

       Min = 2390.9 MWt (−0.65%)

       Max = 2425.6 MWt (+0.66%)

       Average = 2407.6

       Standard deviation = 5.47 MWt

       Average + 3 * standard deviation = 2424.1 MWt

   (b) Based on steam flow from 4000 trials

       Min = 2348.9 MWt (−0.59%)

       Max = 2379.9 MWt (+0.58%)

       Average = 2364.4 MWt

       Standard deviation = 4.53 MWt

       Average + 3 * standard deviation = 2378.0

Further Reading

United States Nuclear Regulatory Commission (2013) *US EPR Application Documents*, http://www.nrc.gov/reactors/new-reactors/design-cert/epr/reports.html#fsar (accessed January 22, 2016).

References

Equations 4.1–4.4, 4.8–4.12, Tables 4.1 and 4.2 from ASME publications are reprinted from ASME PTC 19.5-2013, PTC 6-2014 and MFC-3M-3004 and 2007 Addenda, by permission of the American Society of Mechanical Engineers. All rights reserved. No further copies can be made without written permission.

Rouhani, S. Z. and Axelsson, E. (1970) Calculation of void volume fraction in the subcooled and quality boiling regions. *International Journal of Heat Mass Transfer* **13**, 383–393.
Woldesemayat, M. A. and Ghjar, A. J. (2006) Comparison of void fraction correlations for different flow patterns in horizontal and upward inclined pipes. *International Journal of Multiphase Flow* **33**, 347–370.

Case 5

Two-Phase Hydraulics

Distillation columns are an essential part of petroleum refining, of the processing of natural gas liquids, and of the chemicals industry. In addition to being tall, the columns must be elevated to provide adequate net positive suction head (NPSHR) for product pumps, or control valves, and to provide appropriate differential head for the reboiler located at the bottom of the tower. The reboiler is a natural circulation heat exchanger that provides the tower heat input, driving distillation. The natural circulation through the reboiler is driven by the differential head between the liquid entering and the two-phase fluids leaving. The net positive suction head required by a pump or control valve, or the differential head requirements of the reboiler, will set the height of the tower above grade.

The cost of the tower's foundation, support requirements, and erection costs increase with the tower's elevation above grade. Thus, the tower should be as low to the ground as feasible while meeting the process and equipment requirements of circulation ratio through the reboiler, suction head requirements for rotating equipment and control valves, and providing adequate maintenance access. Failure to provide the required differential head for the reboiler could result in reduced circulation through the exchanger, excess film temperatures that may result in polymerization of hydrocarbons, unsatisfactory product quality or intermittent flow patterns that could overstress the piping, leading to failures, loss of production, and potential fire hazards. Likewise, insufficient net positive suction head for product pumps or control values could lead to increased maintenance or failure of the system to function properly.

Given the foregoing, setting of vessel elevations (Figure 5.1) is a standard element of engineering design for the chemical and process industries. Engineering firms use company standard programs for many of these functions including calculations of the two-phase pressure loss through piping and valves for the reboiler circuits. These programs often incorporate margins to ensure the company's guarantees are met at the least cost, without undue risks.

Case Studies in Mechanical Engineering: Decision Making, Thermodynamics, Fluid Mechanics and Heat Transfer, First Edition. Stuart Sabol.
© 2016 John Wiley & Sons, Ltd. Published 2016 by John Wiley & Sons, Ltd.
Companion website: www.wiley.com/go/sabol/mechanical

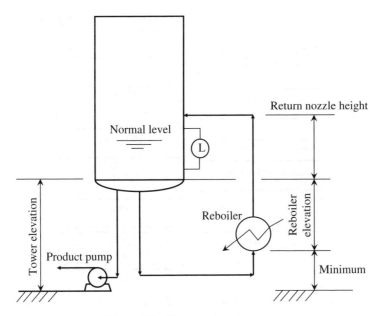

Figure 5.1   Tower setting elevation.

Many times, clients in the petrochemical industry provide their requirements for various design features, including heat exchangers, safety features, equipment redundancy, and two-phase flow hydraulics. In this case, the client for an NGL process plant has required the use of Müller-Steinhagen and Heck correlation with the Moody friction factors for the liquid-only, and vapor-only pressure gradients. The two-phase void fraction is to be determined using Rouhani and Axelsson's 1970 equation I for C reported by Woldesemayat and Ghajar (2006).

These formulations are unfamiliar to your company; but they are provided by the client's specifications. Your company's guarantee related to reboilers and tower elevations will be limited to the correct application of the equations required by the client.

5.1   Background

The presence of two phases, liquid and vapor, in piping systems is a subset of the broader field of multiphase flow. Given the enormous range of engineering applications of multi-phase flow, including transport systems for liquid plus solid, gas plus solid, two-phase liquids, gas plus liquid, condensing and flashing heat transfer, refrigeration, fluid mechanics, power systems, and so forth, there are a number of publications on general and specific topics in multiphase flow.

Two-phase, liquid–gas flow is a complex, nonlinear system, further complicated by phase changes and energy transfer through the system. Models to predict the friction pressure drop in piping systems and the flowing density are abundant. Some models may be general in nature, others are specific to the gas/liquid pair, still others are specific to a flashing liquid in heat transfer, or are written for one flow regime (stratified, annular, slug, churn, horizontal, or vertical flow regimes, etc.)

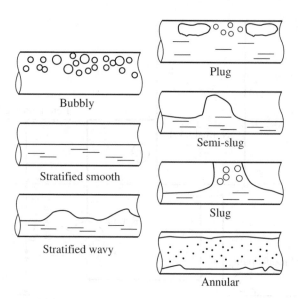

Figure 5.2 Horizontal two-phase flow. *Source*: Reproduced by permission of Scientific & Academic
Publishing

The study and determination of the two-phase flow regime is as varied as that of friction
pressure loss. The mathematical determination of the flow regime is only approximate and is
often not transferrable from one system to another; from an air-water system to refrigerants
for example. Vertical flow patterns are different from those in horizontal pipes, and have
different modeling techniques.

Basically, however, the flow patterns shown in Figure 5.2 (Abdulmouti, 2014) develop in a
predictable manner. In horizontal flows, a "dispersed bubble" flow pattern occurs with a high
volume fraction of liquid and very little vapor. The vapor velocity is determined mostly by the
liquid velocity and there is some gravitational separation between the liquid and vapor.

At the other end of the spectrum, there is the "stratified flow" regime. Here, the volume
fraction of vapor usually exceeds that of the liquid and its velocity is slightly higher. However
the velocities of both phases are rather low – nearly laminar. As the quantity of vapor increases,
the velocity difference between it and the liquid begins to whip up waves on the liquid
surface – "stratified wavy" flow. Gravitational forces are still able to separate most of the
liquid from the vapor.

As liquid volume increases, the waves can reach the top of the pipe creating "slug" or
intermittent flow. Intermittent flow is often divided in to two groups "slug" and "plug" flow.
Plug flow is characterized by a high volume flow of liquid compared to vapor whereas slug
flow is the opposite. Plug flow can develop by increased vapor flow rate from the dispersed
bubble flow regime. As the vapor flow rate increases, bubbles agglomerate to generate the
plug flow characteristic.

As the vapor flow rate increases, the differential velocity between vapor and liquid creates
a channel in the center of the pipe through which the vapor flows. The liquid phase hugs the
internal perimeter of the pipe. The high-velocity vapor usually carries liquid droplets that are
dispersed in the phase. The quantity and size of the droplets depend on the viscosity and

surface tension of the liquid. The flow pattern is referred to as "annular dispersed" (Taitel and Dukler, 1976).

Vertical flow maps are governed by gravitational forces in a different way from horizontal flows. At low vapor flow rates, the vapor tends to break into bubbles – this is known as "bubble flow" – and rise through the liquid. At very low vapor rates, the liquid velocity may dictate the vapor velocity. As the vapor flow rate increases, the bubbles agglomerate to larger bubbles that approach the pipe diameter – "slug" or "plug" flow. Further increases in the vapor flow rate create an "annular dispersed" regime similar to a horizontal flow. Increasing liquid flow in a vertical, annular dispersed regime creates streaks of liquid "wisps" in the core of the annular dispersed flow – "wispy annular dispersed" flow.

When considering vertical two-phase flow, the flowing density becomes an important parameter. Liquid and gas velocities are different, thus the static density of the mixture does not represent the flowing density. As the pressure drop through the piping system increases, the differential velocity increases, and the flow pattern and flowing density change.

A piping system for a two-phase system is usually designed to avoid slug and plug flow regimes. Slugs of liquid striking a bend or entering a vessel can create stresses that were not anticipated. These stresses can lead to cracking of pipe materials, dislodging of vessel internals and potential leaks of hazardous substances.

Velocities that are excessive, creating annular dispersed flow patterns, can often lead to internal erosion of piping. This should be avoided by following accepted industry standards or corporate guidelines and experience. Piping that is subject to a two-phase flow should be inspected on a more frequent basis than piping in single phase applications to ensure erosion is not present and does not lead to a hazardous situation.

5.1.1 Reading Assignment

Read Thorne (2006) and Woldesemayat and Ghajar (2007). Woldesemayat and Ghajar suggested an improved formulation for void fraction by adding a diameter term. What is the error in their formulation?

5.1.2 Müller-Steinhagen and Heck

The Müller-Steinhagen and Heck (MSH) two-phase friction pressure loss equation (Müller-Steinhagen and Heck, 1986) is basically an empirical formulation designed to match a large number of experimental results. Due to uncertainties associated with the calculation of two-phase friction pressure losses, the client applies a safety factor of 20% to the base friction pressure loss when establishing the tower elevation.

The MSH equation is shown in equations (5.1) through (5.4):

$$\left(\frac{dP}{dL}\right)_{fric} = C \cdot (1-x)^{\frac{1}{3}} + Bx^3 \tag{5.1}$$

where:

$$C = A + 2(B-A)x \tag{5.2}$$

In equation (5.1), parameters A and B are the friction pressure losses for all the fluid flowing as a liquid and all the fluid flowing as a gas respectively. These are calculated with equations (5.3) and (5.4) respectively:

$$A = f_L \cdot \frac{G^2}{2\rho_L d} \tag{5.3}$$

$$B = f_G \cdot \frac{G^2}{2\rho_G d} \tag{5.4}$$

Parameter A above can be assumed to be constant through most the piping systems. The author suggests taking the average of parameter B calculated at the pipe inlet and exit.

For laminar flow ($R_d < 2,000$), friction factor is found from equation (5.5):

$$f = \frac{64}{R_d} \tag{5.5}$$

For automated calculations, the Moody friction factor can be approximated with the set of equations 5.6 through 5.9. At Reynolds numbers above about 2000, the friction factor is the lesser of 0.1 and the value from Serghides (1984) equations (5.6) through (5.9):

$$f = \left[\Psi_1 - \frac{\left(\Psi_2 - \Psi_1 \right)^2}{\left(\Psi_3 - 2 * \Psi_2 + \Psi_1 \right)^2} \right]^{-2} \tag{5.6}$$

where:

$$\Psi_1 = -2 \cdot log_{10} \left(\frac{R_r}{3.7} + \frac{12}{R_d} \right) \tag{5.7}$$

$$\Psi_2 = -2 \cdot log_{10} \left(\frac{R_r}{3.7} + \frac{2.51\Psi_1}{R_d} \right) \tag{5.8}$$

$$\Psi_3 = -2 \cdot log_{10} \left(\frac{R_r}{3.7} + \frac{2.51\Psi_2}{R_d} \right) \tag{5.9}$$

5.1.3   Void Fraction

The volume occupied by the gas phase in a two-phase flowing system is defined as the void fraction (ε.) The mean density is calculated from the void fraction in equation (5.10):

$$\rho = \varepsilon \cdot \rho_G + \rho_L \cdot \left(1 - \varepsilon \right) \tag{5.10}$$

Rouhani and Axelsson (Woldesemayat and Ghjar, 2006) suggested using equations (5.11) through (5.13) for void fraction, which is reasonably accurate over a wide range of flow conditions.

Table 5.1 Resistance coefficients.

Valve/fitting	K
90° long radius bend	$14f_T$
Branch "T"	$60f_T$
Gate valve	$8f_T$
Entrance	0.5
Exit	1

Source: Crane Technical Paper 410

$$\varepsilon = \frac{x}{\rho_G}\left[C_0\left(\frac{x}{\rho_G}+\frac{1-x}{\rho_L}\right)+\frac{U_{GM}}{G}\right]^{-1} \tag{5.11}$$

where:

$$U_{GM} = \left(\frac{1.18}{\sqrt{\rho_L}}\right)\left[g\sigma\left(\rho_L-\rho_G\right)\right]^{0.25} \tag{5.12}$$

and:

$$C_0 = 1+0.2\left(1-x\right) \tag{5.13}$$

Piping friction head losses (m/ft) for single-phase systems are calculated with the Darcy–Weisbach equation. The equivalent length of valves and fittings may be determined by Crane (1988). Values of the resistance coefficient of various valves and fittings are shown in Table 5.1 from Crane (1988). The equivalent length of straight pipe may be interchanged with the resistance coefficient using equation (5.15):

Resistance coefficients for valves and fittings are available from a number of sources on the Internet – *The Engineering Toolbox* (http://www.engineeringtoolbox.com/), and pump companies for example. Ding *et al.* (2005) is also an excellent source for elbows, tees, expansions, and reducers.

Piping friction head losses (m/ft) for single-phase systems are calculated with the Darcy equation – equation (5.14):

$$h_f = f\left(\frac{L}{d}\right)\frac{V^2}{2g} \tag{5.14}$$

The equivalent length of valves and fittings may be determined from the Crane Technical Paper No. 410 (Crane, 1988). Values of the resistance coefficient of various valves and fittings are shown in Table 5.1 from Crane (1988). The equivalent length of straight pipe may be interchanged with the resistance coefficient using equation (5.15):

$$K = f\left(\frac{L}{d}\right) \tag{5.15}$$

Table 5.2   Suction specific speed – typical ranges.

Flow units (Q)	NSS range
l/s	4900–7300
m³/s	155–232
gpm	8000–12 000

The fully turbulent friction factor (f_T in equation Table 5.1) is approximated with equation (5.16):

$$f_T = \frac{0.5}{42.944 + 13.55 \ln(d)} + 0.0114 d^{-0.227} \tag{5.16}$$

5.1.4   Pumping Net Positive Suction Head Required

Suction head requirements to prevent cavitation in a product pump may set the elevation for a distillation column. The suction head can be determined from the pump's suction specific speed (N_{SS}) shown in equation (5.17). Technically, N_{SS} is a dimensionless parameter but, for convenience, the gravitational constant is not included and values for the shaft speed (ω), flow volume (Q) and $NPSH_R$, are left in accepted units used by manufacturers or engineers for pump purchase specifications. This practice yields a number for comparison that has typical ranges depending on the units selected. The shaft speed is usually expressed in revolutions per minute. Flow volume may be in m³/s, l/s, or gpm. Net positive suction head is typically either expressed either in meters or feet:

$$N_{SS} = \frac{\omega \sqrt{Q}}{NPSH_R^{0.75}} \tag{5.17}$$

The suction specific speed of a centrifugal pump determines inlet design parameters for the pump, including the inlet flow angle, number of vanes, and inlet guide vanes in the case of very high values for N_{SS}. Low values for N_{SS} permit a wider operating range for the pump without cavitation. Fluid properties can determine the maximum value for N_{SS}. Mixed hydrocarbons can tolerate a higher design value for N_{SS} because there are a number of saturation pressure-causing flash points.

There are no definite design values of N_{SS}, but general rules of thumb for typical ranges are shown in Table 5.2 for the units typically chosen for the volume flow. For this case study, calculate the required suction head based on N_{SS} equal to 7340 l/s, 232 m/s, and 10 000 gpm with a safety margin of 0.6 m (2 ft). By selecting the suction-specific speed, the tower elevation can be established based on the required suction head. This value will then be specified for the manufacturer, who will provide equipment with the necessary design.

5.1.5   Projects

Engineering design projects are generally organized by discipline. Within each discipline there may be various specialties or the specialties may be treated as a separate discipline. Corporate guidelines and a code of conduct should ensure that engineers work within their area of expertise. As the project develops, an initiating discipline will hand off work products to one or more receiving disciplines that advance the design and specifications germane to their specific area of expertise. In this case study, process engineering has handed off an engineering package to the fluid systems engineering discipline, which will contribute to the piping and instrumentation diagrams (P&IDs), set equipment elevations, advance the vessel data sheets, create requirements for control valves, and rotating equipment including pumps, compressors, and fans.

Engineering disciplines may be organized depending of the project size and corporation. An organization chart like the one in Figure 5.3 might be used on a process plant design in the chemical or refining industries where many specialties are required with several discrete processing areas of the plant, each requiring equipment specifically designed for that area.

The overall project of Figure 5.3 would include cost and schedule control, quality assurance, procurement, document control, and support disciplines of legal, human resources, and so forth, as necessary to complete the project.

Projects are accomplished on and iterative basis, beginning with broadly defined objectives, and working toward sufficient detail to allow construction and eventual operation of the new asset, specifically designed to meet a client's or owner's requirements. Large projects may be completed in discrete phases, each terminating with a process gate review that allows senior management to review and verify that the original premise for the project is still viable, economics have not changed, and risks are being addressed, among other attributes. Each project phase would be completed as a separate project. For example, phases prior to final authorization (Front End Loading, FEL) may include: identify an opportunity, select a process or method, refine and optimize the selected process, and complete Front End Engineering Design (FEED). If the project receives approval, then the FEED package would be the launch point for detailed design and construction.

Within a project phase, various disciplines are involved to a greater or lesser extent. Process engineers are typically heavily involved early in the project development but only available in consulting roles during FEED and detailed design. An early development process design package (PDP) may contain only the process flow diagrams (PFD), process description, and basis for design. Fluid systems engineering picks up the early process design and develops the conceptual design for the mechanical systems; producing P&IDs and equipment datasheets for vessels and equipment. A single PFD may results in as many as a dozen P&IDs. Once completed, P&IDs are handed off to the piping designers where a single P&ID may yield a dozen or more piping isometrics. The equipment data sheets work their way to the equipment specialists, who develop purchase specifications that are delivered to procurement specialist for acquisition. Civil and structural engineers require manufacturer data, piping isometrics and stress calculation results for weights and stresses to engineer foundations, supports, and steel structures to support piping and elevated equipment.

As details develop, early documents may be revised. Manufacturer data is added to the P&IDs, as are design information from piping and equipment engineers, and client comments are incorporated. Quality-control inspections may yield necessary changes. Throughout the project, engineering documents are reviewed, revised, reviewed again and finally approved. Pride of authorship gives way to a collaborative effort to provide the client the deliverables required to complete the process within the quality requirements.

Projects may be executed under several structures within a corporation. PMBOK® (Project Management Institute, 2013) describes these as functional, matrix, and projectized. To varying

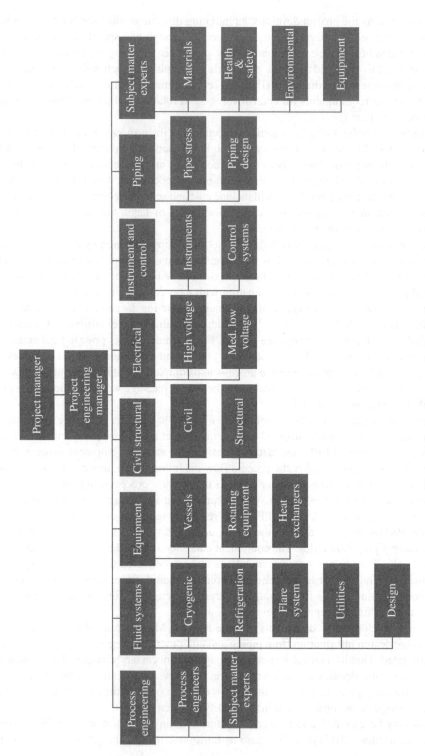

Figure 5.3 Project organization example.

degrees they all share a common trait. When the number of projects diminishes, the staffing diminishes with it. Function and matrix organizations may suffer this to a lesser extent than projectized organizations. In functional or matrix organizations, personnel may be reassigned to a functional department or on another project. An organization assembled for the purpose of a single project can be dismissed at the conclusion of the project. Even with the benefits of function or matrix organizations, companies in the business of executing projects for clients experience reductions in force during periods in which their clients cannot afford to complete more projects.

In this study, you have received the process design from a PFD together with the stream properties as calculated by the process design engineer. You are preparing a calculation to set the elevation of a tower based on the process engineer requirements as outlined in section 5.2.

5.2   Case Study Details

A depropanizer tower separates propane from heavier compounds, mostly butanes and pentane, present in the flow to the tower. The propane exits the top of the tower as a vapor, while the heavier compounds leave the bottom as a liquid (the bottoms.) A portion of the bottoms are boiled in a natural circulation heat exchanger called a "reboiler," which provides the tower's distillation energy. The tower bottoms enter the reboiler from below and a two-phase, gas/liquid flow leaves from above and returns to the tower above the liquid level.

The flow rate through a horizontal reboiler, used in this case, is determined by the elevation difference between the reboiler centerline and the returning elevation to the tower. The greater the difference in elevation, the greater the flow rate. The flow rate has to be sufficient to transfer the energy from the reboiler to the tower without overheating the fluid, and is usually specified by the process or heat exchanger requirements.

The project process engineer has provided the vessel sketch a portion of which is shown in Figure 5.4. The thermodynamic and transport properties of the saturated liquid at the bottom

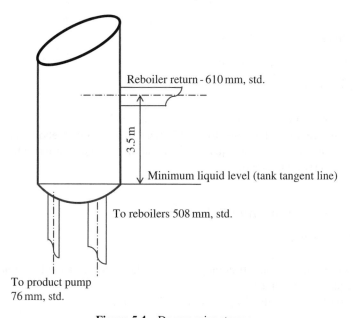

Figure 5.4   Depropanizer tower.

Table 5.3 Depropanizer reboiler parameters.

Parameter	Tower bottoms	Reboiler outlet vapor	Reboiler outlet liquid
Temperature (°C)	41.5	41.6	41.6
Pressure (kPAa)	1380	Calc.	Calc.
Density (kg/m³)	468	Calc.	467.7
Molecular weight	–	45.2	–
Compressibility	–	0.78	–
Viscosity (cP)	0.086	0.009	0.085
Surface tension (dyne/cm)	5.28	–	5.23
Flow (kg/h) (total for dual reboilers)	–	170 700	–

of the tower and the two phase fluid leaving the reboiler, are tabulated in Table 5.3. The reboiler circulation rate should be 1.855 (mass of liquid / mass flow of vapor.) Values shown as "Calc" in Table 5.3 are to be calculated by the engineer setting the tower elevation.

The pressure loss through the exchanger is 1.75 kPa. Working with the piping designer, you have developed the isometric sketch of the reboiler inlet and exit pipelines presented in Figure 5.5. The reboiler discharge lines have been sized to prevent intermittent or slug flow throughout the operating range. For maintenance access, piping must be at least 1 m (3 ft) above grade and the reboiler at least 2 m (6 ft) above grade.

For this case study, assume the electric line frequency is 50 Hz and the pump shaft speed is 1450 rpm. For a 60 Hz application, a pump with the case study capacity typically would have a shaft speed of 1750 rpm. The tower bottoms flow rate to the product pump is 0.95 kg/s. Assume the pump suction centerline is 0.6 m (2 ft) above grade.

5.3   Exercises

5.3.1   Liquid Flow to Reboiler

Calculate the following:

1. Liquid flow rate to the dual reboilers.
2. The total equivalent length of pipe to the reboilers at an assumed reboiler elevation.
3. The friction pressure loss using equation 5.14 and the static pressure rise to the reboiler.
4. The pressure entering the reboiler.

5.3.2   Two-Phase Flow from Reboiler

Calculate the following:

1. The liquid-only and gas-only friction pressure gradients for the two phase lines leaving the reboiler, equations 5.3 and 5.4.
2. The total equivalent length of piping leaving the reboiler.

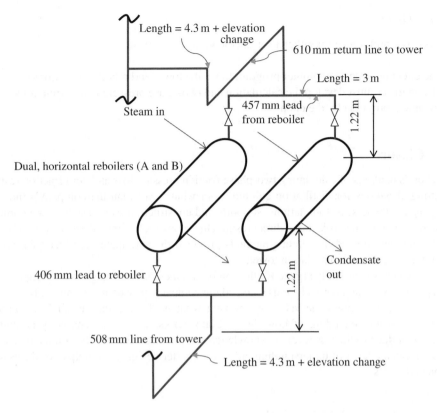

Figure 5.5   Reboiler isometric sketch.

3. The two-phase MSH friction pressure gradient (5.1) and total friction pressure loss in the two-phase return lines to the tower with a 20% margin.
4. The Rohani and Axelsson void fraction: ε (5.11).
5. The two-phase density (5.10) and static pressure loss with 20% margin.
6. The total pressure loss for the two-phase line leaving the reboiler.
7. The elevation of the reboiler centerline such that the pressure leaving the two phase line equals the pressure entering the liquid line from the tower.
8. The elevation of the tower tangent line above grade.

5.3.3   Pump Suction

Calculate the following:

1. The required net positive suction head.
2. The minimum tower elevation above grade to satisfy 1. above.
3. The available net positive suction head at the tower elevation given by section 5.3.2, number 8. above.

5.3.4 Discuss

Discuss your preference for either 1. or 2. below:

1. The client provides a computer program to set the tower and reboiler elevations.
2. The engineer must develop the calculation and two-phase pressure loss correlations from a document provided by the client.

5.4 Closure

The client's methods of calculating two-phase friction pressure loss and void fraction resulted in setting elevations for distillation columns somewhat lower than the company's methods. In aggregate, the cost of foundations, supports and construction were less than the company had anticipated for the project as a direct result. Once constructed, the initial performance tests and evaluations showed the client's methods provided the anticipated product qualities and performance from each distillation tower.

The engineer successfully argued for the company's methods to change. New computer programs, with corporate approval for use, helped the company provide more competitive offers.

This case study shows the value of knowing what is "behind the curtain" of computer programs. An understanding of how the program works, what the technology is, and the source material provides a level of knowledge that exceeds the ability simply to use the program. Such a level of understanding, allows for better designs, and helps avoid misapplication and errors.

5.5 Symbols and Abbreviations

A: friction pressure with the total mass flowing as liquid
B: friction pressure gradient with the total mass flowing as gas
C: found from equation (5.2)
d: pipe inside diameter (L)
f: Moody friction factor (dimensionless)
f_T: Moody friction factor in fully turbulent region
F: force
ft: feet (1 ft = 0.3048 m)
g: acceleration due to gravity (L/t^2)
G: Mass flux ($M/s/L^2$) = $\dfrac{\dot{m}}{pipe\,area}$
gpm: US gallons per minute
h_f: friction head loss (L)
K: resistance coefficient
L: distance or length
M: mass
$\dot{m}$:   mass flow (M/t)
N_{ss}:   suction-specific speed

NPSH$_R$:   net positive suction head required (L)
P: pressure (F/L^2)
Q: flow volume flow rate (L^3/t)
R: Reynolds number $R_d = \rho v d / \mu$
t: time
U_{GM}:   parameter from equation (5.12)
v: specific volume (L^3/M)
V: velocity (L/t)
x: vapor mass fraction
μ: viscosity [M/(Lt)]

Greek Letters

ε: pipe absolute roughness (L); void fraction (dimensionless)
Ψ_1:   parameter from equation (5.7)
Ψ_2:   parameter from equation (5.8)
Ψ_3:   parameter from equation (5.9)
σ: surface tension (F/L)

Subscripts

d: pipe inside diameter
G: gas
L: liquid

5.6 Answer Key

Section 5.1.1. The void fraction is a dimensionless parameter. The suggested formulation results in dimensions of (length)$^{-0.25}$ due to the presence of the diameter in the second term of the denominator of the equation. For laboratory-scale systems, pipe diameters are generally small, and the error from the suggestion may be negligible. However, in commercial-scale facilities the error can be significant.

Section 5.3.1. For reboiler elevation 4.72 m below tower bottom tangent line; pressure drops exclude the reboiler.
  1. 487 348 kg/h.
  2. Segment 1: 2 90° bends, 1 branch T, 1 entrance, 11.7 m straight length at reboiler elevation = −6.2 m from tangent line; total equivalent length = 61.2 m. Segment 2: 1 90° bend, 1 gate valve, 1 exit, 3 m straight length; total equivalent length = 37.9 m.
  3. Friction loss: 1.5 kPa; Static rise: 30 kPa.
  4. Exit pressure: 1.305 MPa.

Section 5.3.2. For a reboiler elevation 4.72 m below tower bottom tangent line, pressure drops including the reboiler:
  5. Liquid only: 6.28E-03 kPa/m; Gas only: 9.04E-02 kPa/m.
  6. Segment 1: 55.5 m, Segment 2: 85.6 m.

7. Segment 1: MSH = 6.04E-2 kPa/m, 4.0 kPa total friction loss. Segment 2: MSH = 5.17E-2 kPa/m, 5.3 kPa total friction loss.
8. $\varepsilon = 0.78$.
9. Segment 1: 127.5 kg/m³ loss = 2 kPa; segment 2: 126.6 5 kg/m³, loss = 10 kPa.
10. Total pressure loss = 23 kPa.
11. Reboiler centerline 4.72 m below tower bottom tangent line.
12. Tower elevation 7.94 m (4.72 + 1.22 + 2).

Section 5.3.3
13. NPSHR = 0.84 m including the safety margin.
14. Minimum tower tangent line elevation 1 m.
15. Tower elevation set by the reboiler. Available NPSH = 6.2 m.

References

Abdulmouti, H. (2014) Bubbly two-phase flow: Part I- characteristics, structures, behaviors and flow patterns. *American Journal of Fluid Dynamics* **4**(4), 194–240.

Crane® (1988) *Flow of Fluids Through Valves, Fitting, and Pipe*. Technical Paper No. 410. Crane Co., Joliet, IL.

Ding, C., Carlson, L., Ellis, C., and Mohseni, O. (2005) *Pressure Loss Coefficients of 6, 8 and 10-inch Steel Pipe Fittings*, University of Minnesota St. Anthony Falls Laboratory, Minneapolis, MN, http://conservancy.umn.edu/bitstream/handle/11299/113368/pr461.pdf?sequence=1(accessed January 24, 2016).

Müller-Steinhagen, H. and Heck, K. (1986) *A Simple Friction Pressure Drop Correlation for Two-Phase Flow in Pipes*, Elsevier, Amsterdam.

Project Management Institute (2013) *Project Management Body of Knowledge*. 5th edn. Project Management Institute, Inc., Newtown Square, PN.

Serghides, T. K. (1984) Estimate friction factor accurately. *Chemical Engineering* **91**(5), 63–64.

Taitel, Y. and Dukler, A. E. (1976) A model for predicting flow regime transitions in horizontal and near horizontal gas-liquid flow. *AIChE Journal* **22**(1), 47–55.

Thorne, J. R. (2006) *Engineering Data Book III*, Wolverine Tube Inc., Decatur, AL, Chapter 13, http://www.wlv.com/wp-content/uploads/2014/06/databook3/data/db3ch13.pdf (accessed January 24, 2016).

Woldesemayat, M. A. and Ghjar, A. J. (2006) Comparison of void fraction correlations for different flow patterns in horizontal and upward inclined pipes. *International Journal of Multiphase Flow* **33**, 347–370.

Case 6

Reliability and Availability

Liquefied natural gas (LNG) provides a means to transport natural gas across great distances without pipeline infrastructure. Those countries and regions of the world without abundant natural-gas resources often import it in the form of LNG via ship. Liquefied natural gas is primarily methane with varying quantities of heavier hydrocarbon impurities, depending on its source and the specific requirements for transportation of the vaporized liquid, cooled to a liquid state. Transportation of the liquid phase via specially designed ships makes it possible to receive a clean-burning energy source competitive with other fossil fuels, renewable energy, and nuclear power sources in regions that lack natural resources.

Countries in the Orient, Japan, and Korea for example, import significant quantities of LNG from around the globe, including Africa, the Middle East, Australia, and North America. Liquefaction facilities compete for market share using price, contract terms, LNG quality, and so forth, to secure supply contracts with perspective buyers.

Your company has been commissioned to design and build an LNG import terminal. The terminal will involve dredging the harbor to accommodate the large LNG tankers, a new berthing facility, offloading equipment to remove the LNG from the tankers, LNG storage facilities, regasification equipment to convert LNG to natural gas, compression equipment to supply the regasified LNG into transcontinental pipelines, and 50 km of pipeline laterals to connect the facility to existing transcontinental pipelines. Stakeholders in the project include the national government, the local harbor master, local communities, shipping companies, suppliers of LNG, and natural gas customers.

Compression of the regasified LNG into the pipelines will require a significant amount of power. Liquefied natural gas is stored at atmospheric pressure in large insulated tanks. From the storage pressure, the natural gas must be compressed to approximately 6.9 MPa to meet the pipeline requirements. As the import terminal is located in a remote location, a new power

Case Studies in Mechanical Engineering: Decision Making, Thermodynamics, Fluid Mechanics and Heat Transfer, First Edition. Stuart Sabol.
© 2016 John Wiley & Sons, Ltd. Published 2016 by John Wiley & Sons, Ltd.
Companion website: www.wiley.com/go/sabol/mechanical

plant is included in the project. Waste heat from the power plant will supply the energy required to regasify the LNG. Excess electricity produced will supply the local market as a byproduct.

The project has a requirement to meet 98% regasification availability on an annual basis. That is, the regasification facilities should be able to provide the full design capacity of natural gas to the pipeline for all but 175 hours during a year – about 1 week. Unexpected outages or curtailments in production would result in penalties itemized in sales and delivery contracts. To meet the project's economic targets, the waste heat supply from the power plant should have an availability greater than the 98% target.

Prior to submitting a proposal for the power plant and waste heat supply, a verified calculation of the power system heat supply availability is necessary.

6.1   Background

Availability and reliability are often interchanged. However, "reliability" is a term that most often refers to the time an equipment item or system is operating at 100% output, neglecting the time the unit must be removed from service for scheduled maintenance. Reliability, given by equation (6.1), is a representation of how well the unit, or process will perform between scheduled maintenance outages. The period over which the reliability is calculated is much greater than the planned maintenance hours, usually 1 year:

$$R = \frac{Operating\,hours - Equivalent\,outage\,hours}{Period\,hours - Planned\,maintenance\,hours} \qquad (6.1)$$

where:
  R: reliability factor;
  Operating hours: total operating hours in the period;
  Equivalent outage hours: see equation (6.2);
  Period hours: total hours in measurement period;
  Maintenance hours: hours removed from service to perform scheduled maintenance.

$$Equivalent\,outage\,hours = \frac{Forced\,partial\,outage\,hours * Size\,of\,reduction}{Maximum\,dependable\,capability} \qquad (6.2)$$

(based on Appendix F, NERC Generating Availability Data System)

Availability, defined in equation (6.3), includes the maintenance hours and is, therefore, a representation of overall production capability over a given period. While reliability provides an indication of how well the process was designed, and how effective maintenance is, availability is necessary for the economic evaluations of the total system.

$$A = \frac{Operating\,hours - Equivalent\,outage\,hours}{Period\,hours} \qquad (6.3)$$

Where A: availability.

An availability, or reliability, calculation is a statistical analysis of a system that yields the most likely fraction of time the system will be capable of providing a desired output.

Calculations may be deterministic or stochastic. Deterministic calculations provide the most likely system availability whereas stochastic models are able to provide a range of likely operations. As random variables are used in stochastic models, they will not produce the same result for each calculation. The two, when properly employed, yield similar results for the most likely value for availability.

6.1.1   Models

Availability / reliability models generally start with a block flow diagram with details appropriate for the analysis, and knowledge at the time of the calculation. Throughout a project development, the model may be modified to include greater detail and specifics regarding the availability of individual items of equipment or process groups as data is verified during the project.

The simple two-process series model shown in Figure 6.1 requires Process 1 and Process 2 to be in service in order to produce the desired output. The determinist calculation of reliability for a series of processes (Process 1, and Process 2, Process 3, etc.) is the multiple of the reliabilities of each individual process. In the case of Figure 6.1, the combined reliability of the two process groups would be 0.95 * 0.92, or 87.4%. In general, the calculation of system reliability for a series of processes required to produce an output is shown by equation (6.4):

$$R_s = R_1 \cdot R_1 \cdot \ldots R_n \tag{6.4}$$

where:
R_s = system reliability;
R_1 through R_n: reliability of series components from 1 to n.

The parallel model of Figure 6.2 shows two 100% capacity processes, each capable of providing the required output. Assuming that the impact of switching reliability is negligible, the combined reliability of a generic parallel system can be calculated using equation (6.5):

$$R_s = 1 - \left[(1 - R_1)(1 - R_2) \ldots (1 - R_n) \right] \tag{6.5}$$

where:
R_s = system reliability;
R_1 through R_n: reliability of parallel components from 1 to n.

For the simple two parallel processes of Figure 6.2, the combined reliability would be 99.75%. As demonstrated, adding redundant equipment in parallel significantly improves reliability at a predictable added capital cost. The value of the additional capital can be determined through an economic analysis by comparing the output and earnings of the option compared to alternatives.

Figure 6.1   Series reliability model.

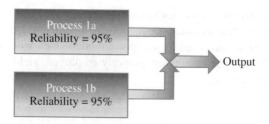

Figure 6.2 Parallel process reliability model.

Another method of improving reliability over a system with a single component is to construct a parallel system with a set of processes, each capable of producing an equal portion of the output. For example, three 50% capacity pumps (3 × 50%), or four 33% capacity units (4 × 33%). Such a system is often abbreviated as an "*x* of *n*" system – that is, "*x*" of the total number "*n*" are required to meet 100% output. The fraction of time the system would be capable of meeting 100% output can be calculated with the binomial distribution, equation (6.6):

$$R_s = \sum_{r=0}^{k} \frac{n!}{r!(n-r)!} p^r (1-p)^{(n-r)}$$

(6.6)

where:
R_s: system reliability;
k: maximum number of failures allowed while meeting 100% output;
n: total number of parallel equally sized units in the system;
r: the number of units out of service;
p: $1 - R$;
R: reliability of each unit.

The binomial distribution yields the probability of achieving 100% load with the components installed. Figure 6.3, containing five units of one-third capacity each, requires three of the five to produce 100% design output. Two units can fail and the system will continue to produce at the design output. Each unit is identical and has a reliability of 90%. From the binomial distribution, the system can produce 100% output 99.14% of the time.

The binomial evaluation, as implied by its name, accounts for the time the system can meet a specific output. Any scenario not meeting the exact criteria is excluded from the result. In the case of Figure 6.3, for example, two units in service would produce two-thirds of the required output, a partial load output that contributes to the overall system reliability but is not included in the previous example. Again using the binomial distribution, the percentage of time that exactly two units are in service (three failures) is 7.3%. Their output, 66.7%, added to the previous result, yields an overall reliability of 99.68%.

Practical systems include combinations of parallel and series process, Figure 6.4 for example. The reliability of such systems may be solved by treating groups of parallel processes as a series component in the overall block flow diagram. Figure 6.4 shows that the components with the lowest reliability are installed as a group of 3 × 50% parallel units; the reliability of which, from the binomial distribution is 97.2%. From equation (6.4), the system reliability of Figure 6.4 would be 93.4%.

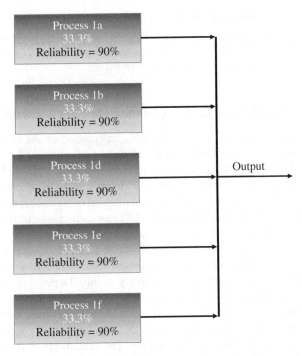

Figure 6.3 Parallel system: x of n required.

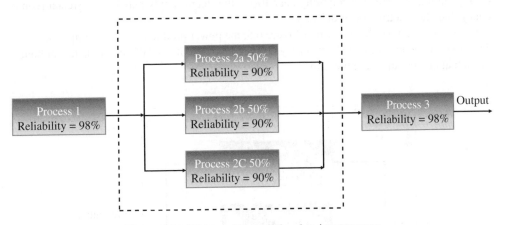

Figure 6.4 System with parallel and series processes.

An aspect of a series system is that a single component in the series can remove the entire system from service. In Figure 6.4, failures in processes 1 or 3 would each be considered a "single point failure" that could remove the system from service regardless of how reliable the parallel system of Process 2 is. Increasing the number of series components in a system, increases the risk of a complete shutdown. When designing systems that require high reliability, engineers often employ redundant equipment installed in parallel to mitigate a potential failure of the complete system.

In noncritical systems, where equipment can be repaired quickly, a redundant spare may be located in the warehouse. For example, a small capacity pump that feeds a storage tank may have a redundant spare in the warehouse. If the pump fails, tank capacity can provide continuous service while the pump is either replaced or repaired.

6.1.2 Availability: Planned and Unplanned Outages – Parallel Systems

Economic considerations of real systems require a calculation of the overall system availability. These calculations are complicated by planned and unplanned outages. In parallel systems, unplanned outages can occur on operating units during a planned outage of one of the parallel units. In series or parallel systems, random unplanned events on a single unit impact availability only during the period between planned outages of that unit.

An availability calculation can consider many different scenarios of planned and unplanned events. Simultaneous outages, such as two unplanned events or an unplanned event during a planned outage, may have a low probability. In most cases, industrial equipment has an unplanned outage rate less than 5%. Therefore, the probability of a random failure of two or more units simultaneously becomes diminishingly small. Depending on the quality of information and assumptions, the overall availability calculation may ignore two or more simultaneous random events due to their significance and impact on the calculation.

In addition, a single process may have more than one output. A cogeneration plant, for example, provides electricity and steam for heating. The availability of the two products may be different depending on the requirements and design of the system. As the host facility for a CHP plant requires steam availability near 100%, the steam supply may have a greater redundancy than the electric power delivery.

Considering the parallel system of Figure 6.5, the power production (P) is composed of 3 × 33.3% components while the heat supply (H) is comprised of 3 × 50% components, although the actual process units are the same.

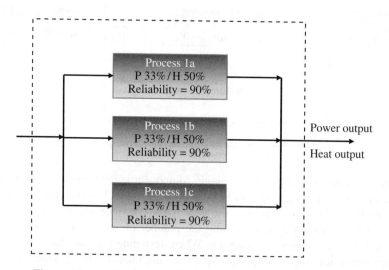

Figure 6.5 Parallel system with two outputs ("P" power, "H" heat).

When calculating the availability for a system with planned and unplanned outages, each type and combination of outages is considered individually until either every combination is considered, or further calculations yield insignificant impacts on the overall system availability.

Planned outages are not random events, so the outage durations for the units within a system are additive. Planning practices generally overlap the maintenance outages for series components to reduce the overall impact on the system availability. The component with the longest critical path outage would determine the duration of nonproductive time. For a parallel system, the total period of planned outages would be the sum of the planned outage durations for each of the units in parallel, less any overlapping outages such as a forced outage occurring during a planned maintenance outage as shown in equation (6.7):

$$POR_s = \sum_{i=1}^{i=n} POR_i - P_{P+F} \qquad (6.7)$$

where:

POR: planned outage rate (planned outage hours / period hours);
P_{P+F}: probability of a planned outage and a simultaneous forced outage;
s: system comprised of units i through n.
i: POR or unit I;
n: number of parallel units;

The probability of a forced or unplanned outage for units in a parallel system is given by the binomial distribution equation modified for the fraction of time consumed by planned outages as shown in equation (6.8):

$$P_F = \frac{n!}{r!(n-r)!} \cdot \left[FOR \cdot (1 - POR_s - P_{P+F}) \right]^r \cdot (1 - FOR)^{(n-r)} \qquad (6.8)$$

where:

P_F: probability of a forced outage;
P_{P+F}: probability of a planned and simultaneous forced outage;
FOR: forced outage rate.

An unplanned event may occur during a planned outage on a parallel process. The probability of such an event can be calculated with equation (6.9). As mentioned above, there is the possibility of experiencing more than one forced outage in a parallel system while one of the units is out of service for scheduled maintenance. However, the probability of such an event is so low that it is very often ignored.

$$P_{P+F} = n \left(\frac{(n-1)!}{r!(n-r-1)!} \right) \cdot FOR^r \cdot POR \cdot (1 - FOR)^{(n-1-r)} \qquad (6.9)$$

The product of FOR and POR is generally very low. However, a facility with numerous parallel processes yields a high number of possible combinations of units out of service. In operating facilities, unplanned outages occasionally occur during planned maintenance on parallel processes. These multiple outage events can be an important component in an availability calculation. As an example, given a planned outage rate of 2%, the probability of experiencing a forced outage during planned maintenance on one of the processes of Figure 6.3 would be almost 3%.

After calculating the probabilities for each likely outage scenario, the overall system availability is calculated by summing the products of the lost output for each scenario and its probability and subtracting the sum from 1, as in equation (6.10).

$$A = 1 - \sum_{i=1}^{n} P_i \left(lost\, output\, fraction \right)_i \tag{6.10}$$

where lost output fraction = fraction of output lost due to event i.

6.1.3   Series and Parallel Processes

A series/parallel system, such as the one shown in Figure 6.4, adds additional complexity in availability calculations. Avoiding double counting of outage periods requires eliminating overlap of outages that may occur at the same time. In the same way equation (6.7) includes the term $(1 - POR_S - POR_S FOR)$, calculating the probability of a forced outage of a unit within a parallel block that is in series with other processes needs to exclude the overlapping outages in the series components.

For example, when calculating the probability of a forced outage of the parallel components of Process 2 in Figure 6.4 using equation (6.7), the planned and unplanned outage rates of Processes 1 and 3 should be subtracted as well. The reason being that a forced outage within Process 2 could not occur if either Process 1 or 3 had caused the entire system to shut down.

Likewise, an unplanned event in Process 3 would not be possible if either Process 1 or 2 had already removed the system from service. Therefore, the period of time that Processes 1 and 2 are calculated to be out of service is excluded from the unplanned events of Process 3.

The foregoing can be summarized by modifying Equation (6.8) as follows in equation (6.11):

$$P_F = \frac{n!}{r!(n-r)!} \cdot \left(FOR \cdot (1 - \mathcal{O}) \right)^r \cdot (1 - FOR)^{(n-r)} \tag{6.11}$$

Equation (6.9) would be modified as shown in equation (6.12)

$$P_{P+F} = n \left(\frac{(n-1)!}{r!(n-r-1)!} \right) \cdot FOR^r \cdot POR \cdot (1 - FOR)^{(n-1-r)} \cdot (1 - \mathcal{O}) \tag{6.12}$$

where $\mathcal{O}$ is the sum of probabilities of overlapping outages in the series/parallel system. This value changes depending on the event probability being calculated.

As the number of parallel and series components increases, the complexity of the deterministic model also increases. Through the development of a project, the availability / reliability calculation can become difficult to understand, follow, and check. Therefore, specific software-modeling packages, or other means of determining system reliability, are often employed.

After calculating the probabilities of the various scenarios, multiply the lost capacity of the facility by the probability of each, and sum to find the total losses over a period. Subtracting the total losses from unity yields the overall facility availability.

Example 6.1

Let the planned outage rate of the components in Figure 6.5 be 6% and the forced outage rate be 3%. Calculate the overall availability factor for power delivery considering the following scenarios:

- planned outages;
- one planned and one forced outage;
- one forced outage.

From equation (6.7), the planned outage rate for the system would be:

$$POR_s = 3*0.06 - P_{P+F} = 15.32\%$$

Equation (6.9) for the probability of a forced outage during a planned outage yields:

$$P_{P+F} = 6 \cdot 0.03 \cdot POR_s \cdot (1-0.03)^{3-1-1} = 2.68\%$$

Equation (6.8) for the probability of a single forced outage yields:

$$P_F = 3 \cdot \left\{ 0.03 \cdot (1 - POR_s - P_{P+F}) \right\}^1 \cdot (1-0.03)^{3-1} = 7.73\%$$

The solution to the set of equations shown above is iterative.

Lost capacity for a single outage (one planned or one forced) would be 33.33%. For two outages, it is 66.67%. Summing the multiple of each of the probabilities and the respective lost capacity from the above yields 9.21% and an availability of 90.79%.

6.1.4   Stochastic Models

Stochastic models calculate reliability and availability through a series of trials with random numbers to mimic the random nature of unplanned events. A Monte Carlo simulation is one type of stochastic modeling.

In a Monte Carlo calculation, start by setting the output of a process equal to 1 unless (i) there is a scheduled outage period or (ii) a random number is less than or equal to the forced outage rate for the specific process, in which case the output would be zero. For parallel processes, add the output of each unit within the process with a maximum of 1 to find the fraction of output available for the parallel block. Once calculated, the parallel process block can be treated as a single process unit in a series calculation.

Calculate the availability of series components in the same way as a single process in a parallel set of processes. There may be planned maintenance events or random unplanned events determined by a random variable. Multiply the fractional output of series components to yield the net output of the complete system.

Repeat trials until there is a reasonable convergence on the calculated availability. The average availability may be calculated by summing the results of each trial and dividing by the total number of trials.

Complicated systems can be modeled, and calculated very quickly with a Visual Basic program embedded in a spreadsheet. As mentioned earlier, each calculation of a stochastic model will yield a different answer. However, given a large number of trials, the results should be fairly close to one another.

6.1.5   Reading

Read the online technical paper *Reliability Engineering Principals for the Plant Engineer,* by Drew Troyer, available at http://www.reliableplant.com/Read/18693/reliability-engineering-plant (accessed January 26, 2016). Answer the following questions related to the article:

1. The most prevalent reliability model is:
   (a) The bathtub curve.
   (b) The exponential distribution.
   (c) The Gaussian or normal distribution.
   (d) The beta distribution.
2. The Weibull distribution;
   (a) Can mimic the bathtub curve.
   (b) Was written by a Swedish mathematician.
   (c) Relies on field data.
   (d) All of the above.
3. Is the following statement true or false? "When failure modes are lumped together, there appear to be random failures. However, when the failure modes are analyzed a time dependency normally appears."

6.1.6   Applicability

The methods discussed above for the calculation of reliability and availability rely on planned maintenance and random unplanned events that remove equipment or processes from service. Not all unplanned events are random. In fact many are traceable to a systematic cause such as material selection, design defect, misapplication of equipment, improper operations or maintenance, or normal wear and tear. For example: continued use of lubricating oil that has a high concentration of oxidation products and has exhausted its useful life in a set of parallel processes can lead to similar failures, all at about the same time. Such a systematic failure mechanism would result in a sequence of failures that would not be predicted by statistical methods based on random events outlined above.

Human error is in some respects random; however, many mistakes can be mitigated or corrected through proper training, improved procedures, quality assurance practices, quality-control measurements, and so forth. As such, many human errors can be traced to a systematic cause.

Most, if not all, causes of equipment failure can be determined through rigorous root-cause analysis. When followed by corrective actions, systematic sources of unplanned events can be mitigated or completely eliminated through better design, operating practices, or other methods determined through the root-cause analysis process. Root-cause analysis tools and techniques, such as Six Sigma, Ishikawa diagrams, and fault trees are an important aspect of engineering, but are beyond the scope of this case study. The interested student should read Gano (2007).

6.2   Case Study Details

6.2.1   Initial Block Flow Diagram

The initial block flow diagram for the CHP plant associated with the LNG regasification terminal is shown in Figure 6.6. Here there are three gas turbine/HRSG trains, each providing an equal portion of the power output and half of the steam required for the heat needed to vaporize the LNG. Supplemental firing in the HRSGs allows for higher steam temperatures and enables the units to maintain a constant steam-turbine output throughout the year as ambient temperatures affect the flow through the gas turbines.

Steam from each HRSG combines and enters a single steam-turbine generator. The latent heat of the steam-turbine exhaust, normally discarded to the atmosphere, provides useful energy to vaporize LNG. Only two gas turbines are required to vaporize the LNG flow rate, so

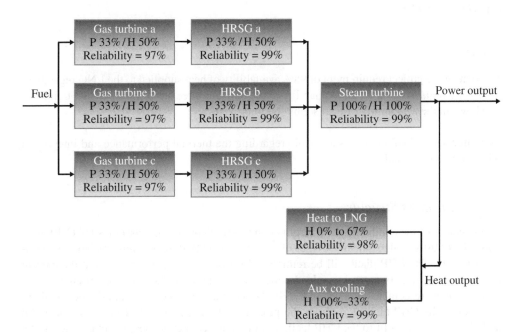

Figure 6.6   Initial CHP block flow diagram with power (P) and heat (H) outputs.

an auxiliary cooling system supplements the cooling capability for the steam-turbine exhaust provided by vaporizing LNG.

Under normal operation, one-third of the total heat rejection from the steam turbines is provided by the auxiliary cooling system. The auxiliary cooling system is capable of providing the full heat rejection requirement when the LNG regasification terminal is out of service.

Normal full-load operation of the LNG regasification terminal requires only two gas turbine/HRSG trains to be in service. A single gas turbine provides enough steam to the steam turbine generator for half of the design LNG flow.

Power output is the sum of the gas-turbine and steam-turbine outputs. If a gas turbine is out of service, the steam turbine output will drop by one-third.

For the project, the gas turbines would be aero-derivative units that require a maintenance outage every 4200 equivalent operating hours. After 29 400 equivalent hours, each gas turbine requires a major inspection. The outage durations can be reduced significantly with a rotor subscription service provided by the manufacturer. In such services, the entire rotor is swapped for a subscription rotor during the outage. The manufacturer repairs the rotor in its shop and makes it available to another service subscriber.

The outage pattern repeats until the end of the project life. In addition to the gas-turbine maintenance, there are four 8-hour, offline water washes conducted on each gas turbine compressor each year. Once every 6 years, the steam turbine must be removed from service for 4 weeks for a maintenance inspection. HRSG inspections are conducted with the gas-turbine outages.

On average, the gas-turbine planned outage rate (POR) is 1.837%. The reliability of a gas turbine is expected to be 97% and the reliabilities of an HRSG and the steam turbine are expected to be 99%.

Exercise

1. To meet the project requirement of 98% availability of heat supplied to the LNG regasification terminal, what must the availability of the block of three gas turbine / HRSG trains be?
2. How can the availability of heat supply be improved?

The information in Table 6.1 is available regarding the thermal performance and integration with the LNG terminal.

6.2.2 Business Structure

The structure of the combined LNG terminal and CHP plant business assumes the CHP plant will operate as an independent provider of electricity and heat energy required for vaporization of LNG. The CHP plant will be reimbursed for the heat energy provided at the current cost of pipeline quality natural gas. Under these terms, the LNG facility receives energy as if natural gas were combusted and used at 100% thermal efficiency for vaporization without emissions. The CHP facility, receives the price of natural gas for low-level waste heat. An added secondary benefit to the CHP facility is that the steam cycle heat rejection occurs at a constant temperature throughout the year.

Table 6.1 CHP/LNG terminal integration.

Parameter	Value
Gas turbine net output each (MW)	49.2
Steam turbine net output (MW)	73.3
CHP auxiliary power (MW)	9.9
CHP net output (MW)	211
Steam-turbine exhaust flow (kg/s)	60.8
Steam-turbine exhaust enthalpy (kJ/kg)	2,239
Steam-turbine exhaust pressure (kPa)	2.51
Gas-turbine fuel (GJ/h) HHV	1,482
HRSG supplemental firing (GJ/h)	225.3
LNG heat requirement (kJ/kg)	879.1

Additionally, the CHP plant receives a price for electricity from the LNG terminal that is above the market rate. The higher than market rate is necessary to:

- compensate the CHP plant for maintaining a high electrical output in all market conditions, even when the retail price for power is below its cost;
- compensate the CHP plant for providing the equipment and systems that deliver the hot water supply to the LNG facility for vaporization; and
- amortize the capital cost for extra redundancy inside the CHP plant required to achieve high availability of the heat supply.

Together, the price for the heat and power supplied to the LNG regasification terminal benefits both parties. The LNG terminal pays a reasonable price for heat and power without having to finance capital investments, and avoids environmental emissions. The LNG terminal owner is provided with highly reliable utilities and can concentrate on its primary business without having to develop talent in a noncore business venture. The CHP plant sells waste heat at the value of natural gas and receives payment for additional capital investments that were necessary over and above that which is normally required by a typical power generator.

Given the foregoing, the net effective heat rate (NEHR) of the CHP plant is given by equation (6.13)

$$NEHR = \frac{Q - 0.8791 * LNG}{Net\ electric\ output} \tag{6.13}$$

where:
NEHR: net effective heat rate (GJ/MWh);
Q: total fuel requirement for the CHP plant (HHV) GJ/h;
LNG: Vaporized LNG flow rate (te/h);
Net electric output: net electric output from the CHP plant (MWh).

The NEHR may be calculated hourly as shown in equation (6.13) or on a monthly basis with Q, LNG, and electric output determined for the entire month. Further negotiations will determine the exact methods and calculations. Bonuses and penalties for availability guarantees will also be negotiated to ensure both parties are adequately incentivized to properly maintain the facilities throughout the economic life of the project.

An additional benefit to the CHP plant is that the NEHR is lower (the efficiency is higher) than would be possible with a pure power generating facility. This allows the facility to remain competitive in difficult economic conditions, and as advances in technology create ever more efficient power generating facilities.

6.2.3   Modified Block Flow Diagram

Evaluations of the initial block flow diagram revealed that the steam turbine was a single-point failure of the waste heat supply to the LNG terminal. Without a backup, the steam turbine could remove the entire CHP facility and LNG regasification terminal from service. While the calculated system heat-supply availability was greater than 98% for the first five years, the sixth year would suffer due to the planned steam-turbine maintenance. The design team had also expected a low forced outage rate for the steam turbine. If actual operation was even slightly worse, the overall LNG terminal availability could fall significantly below the target of 98%.

A steam bypass system was therefore designed around the steam turbine to permit operation of the gas turbine / HRSG trains without the steam turbine. Heat for the LNG regasification terminal would be provided by the steam generated in the HRSGs, which would be cooled with condensate to prevent systems that normally operate with steam turbine exhaust from overheating. The system included redundant pumps for cooling, highly reliable instrumentation, and valves yielding a robust system with high reliability. As such, the development sponsor agreed that it was unnecessary to consider that the bypass system would not operate in the event of an unplanned outage of the steam turbine.

In order to conduct a steam-turbine maintenance outage, a complete shutdown of the CHP plant was required to isolate the steam turbine from the condenser. The outage was estimated at 2 days. The reverse process of removing the isolation would require another 2-day outage of the CHP plant prior to placing the steam turbine back in service. These outages would be scheduled with the LNG facility to reduce economic penalties but the system was still required to meet an overall availability of 98% or better.

The modified block flow diagram is shown in Figure 6.7.

The auxiliary cooling system was to be a wet cooling tower comprising four parallel cells with three 50% capacity cooling water pumps, each component being highly reliable. The overall auxiliary cooling water system would therefore have a reliability of approximately 99.9%. As with the steam bypass system, the project team and sponsor agreed that it was unnecessary to consider a complete or significant failure of the auxiliary cooling system together with another simultaneous unplanned outage. The reliability impact of the auxiliary cooling system could therefore be ignored in the overall system waste heat and power availability calculations.

6.2.4   Other Considerations

Natural gas from the LNG regasification terminal would be connected to multiple interstate pipelines. In the event that the LNG terminal would be out of service, natural gas could be provided to the CHP from the interstate systems, which had established very

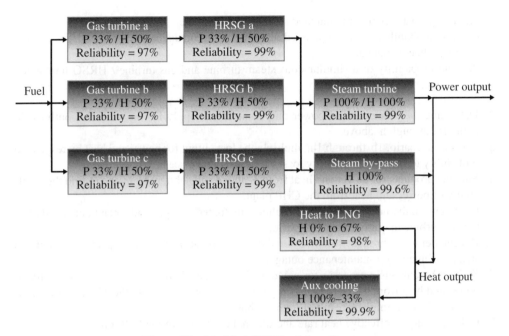

Figure 6.7 Modified CHP block flow diagram.

nearly 100% reliability. The design team therefore felt that there was no need to consider a complete fuel outage.

In the event of a failure of the local electric power grid, the CHP could operate at a reduced load to provide power for the operation of the LNG terminal and regasification facilities. Further economic analysis would be required to determine if the CHP plant needed capabilities to start in the absence of the local power grid. Initial discussions and data on the grid's reliability indicated that the "black start" capability would not be necessary.

6.2.5   Exercises

Complete the following:

1. Set up a spreadsheet to calculate the availability of power and heat provided by the CHP plant discussed above. Neglecting the sixth operating year, which includes a steam turbine outage, calculate the following for the gas turbine / HRSG block:
   (a) The effective forced outage rate for one gas-turbine / HRSG train.
   (b) The system planned outage rate neglecting outages coinciding with a forced outage.
   (c) The probability of a single unplanned outage.
   (d) The probability of experiencing an unplanned outage during planned maintenance of a gas turbine.
   (e) The probability of two simultaneous unplanned gas turbine / HRSG outages.

(f) The probability of an unplanned steam-turbine outage.

(g) The probability of an unplanned stream turbine outage with a planned gas-turbine maintenance outage.

(h) The probability of a simultaneous steam turbine and gas turbine / HRSG unplanned outage.

2. Determine the percentage of power output and heat supply for the scenarios outlined in items 1b. through h. above.

3. For each scenario, 1b. through 1h., multiply the lost output for power and heat by respective probability to find the availability losses for each outage scenario.

4. Sum the products for power and heat from 3. above and subtract from one to find the availability of power and heat from the CHP plant.

5. Comment on the need to consider simultaneous forced outages and forced outages during the gas turbine major maintenance periods.

6. Repeat items 1. through 4. above for the sixth operating year, taking into account the steam-turbine major maintenance outage.

7. Write a Visual Basic program for a Monte Carlo analysis of the expected availability of power and heat from the CHP plant for years 1 through 5. Compare the Monte Carlo analysis and the deterministic result of item 4. above.

8. Calculate the net effective heat rate and thermal efficiency of the CHP plant.

6.3   Closure

This case study develops an understanding of the deterministic and stochastic calculation of reliability and availability. This can be employed in a wide variety of applications including design, project development, plant engineering, consulting, project management, quality control, and information technology. The author maintains these tools with others for presentations, and specialty calculations that can be readily accessed.

The development project referenced above was a novel use of waste heat from a steam turbine exhaust to vaporize LNG. The low-level waste heat from the power cycle was adequate for the cryogenic liquid, and improved the economics of the power generating facility. The application improved the environmental impact of the proposed facility by eliminating the need for seawater extraction, and avoided disturbing sensitive offshore ecosystems. To gain acceptance of the concept by the project sponsor, the development engineer had to clearly establish the availability of the heat supply for the LNG. The tools outlined in this case study were an integral part of that success.

6.4   Symbols and Abbreviations

R: reliability, equation 6.1
A: availability, equation 6.2
CHP: combined heat and power
FOR: forced outage rate (unplanned outage hours / period hours)

HRSG: heat recovery steam generator
k: minimum number of operating units to meet 100% output
LNG: liquefied natural gas
n: total number of units
NEHR: net effective heat rate (GJ/MWh), usually defined by a power purchase agreement
$\mathcal{O}$:   the sum of probabilities of other overlapping relevant outages in the series / parallel system
p: $1 - R$, probability of an unplanned outage on a single unit or process
P: probability
POR: planned outage rate (planned outage hours / period hours)
Q: heat energy (GJ/h)
r: number of units failed

6.5   Answer Key

Section 6.1.5

1. B
2. D
3. True

Section 6.2.1

1. Meeting an overall availability of 98% with a steam turbine in series with the gas turbine/ HRSG block requires the gas turbine block to have an availability exceeding 99%.
2. A component in parallel with the steam turbine that would allow the gas turbine block to operate without the stream turbine would improve overall reliability.

Section 6.2.5

(a) 3.97%
(b) 5.052%
(c) 10.182%
(d) 0.414%
(e) 0.394%
(f) 0.838%
(g) 0.044%
(h) 0.095%

Outage scenario	Probability (%)	Power capacity (%)	Waste heat capacity (%)
1 planned outage	5.052	67	100
1 forced outage	10.182	67	100
1 planned and 1 forced	0.414	33	50
2 GTG/HRSG forced outages	0.394	33	50
STG outage only	0.838	66.5	100
STG outage + GTG planned	0.044	44	100
STG outage + GTG forced	0.095	44	100

3. and 4.

Outage scenario	Lost power capacity (%)	Lost waste heat capacity (%)
1 planned outage	1.684	0.000
1 forced outage	3.394	0.000
1 planned and 1 forced	0.276	0.207
2 GTG/HRSG forced outages	0.262	0.201
STG outage only	0.281	0.000
STG outage and GTG planned	0.025	0
STG outage and GTG forced	0.053	0
Availability (power/heat)	94.03	99.59

5. The assumed forced outage rates for the gas turbine, HRSG and steam turbine have a single significant digit; therefore, outages with a probability less than 0.5% are likely insignificant in the overall calculation of availability. However, the sum of one planned plus one forced outage, and two gas turbine forced outages is greater than 0.5%; therefore, the likelihood of two units being out of service simultaneously is significant and should be considered in the project planning. The other simultaneous outages involving an unplanned steam turbine outage with either a planned or unplanned gas turbine outage are insignificant given the level of accuracy of the forced outage rates.

6. As shown by the table below, the waste heat available was expected to exceed the project requirements with the steam by-pass line installed.

Power availability, year 6			
Power availability	Probability (%)	Power capacity (%)	Lost power (%)
48 weeks			
1 planned outage	5.052	67	1.554
1 forced outage	10.182	67	3.074
1 planned and 1 forced	0.414	33	0.255
2 forced outages	0.394	33	0.242
STG outage only	0.838	0.665	0.769
STG outage + GTG planned	0.044	44	0.023
STG outage + GTG forced	0.095	44	0.049
4 weeks			
1 planned outage	5.096	44	0.218
1 forced outage	10.182	44	0.436
1 planned + 1 forced	0.414	22	0.025
2 forced outages	0.393	22	0.024
96 hour outage	1.096	0	0.084
Total lost power			**6.752**
Power availability: GT/HRSG Train			**93.25**
Steam availability year 6		Heat capacity	Lost heat
96-hour outage	1.096	0	0.084
Normal steam availability			99.592
Steam availability year 6			**99.508**

7. Results of the deterministic calculations and the Monte Carlo simulation agree.

Percent of total power output	Deterministic (%)	Monte Carlo (%)
0	Not available	0
22	Not available	0
33	0.808	0.879
44	0.139	0.171
65	0.838	0.890
66.67	15.23	15.27
100	83.08	82.79
Power availability	94.03	93.94 (five calculations of 8760 trials each)
0	Not available	0
50	0.816	0.879
100	99.184	99.121
Heat availability	99.59	99.59 (five calculations of 8760 trials each)

8. Net effective heat rate = 6.66 GJ/MWh (54% thermal efficiency).

Reference

Gano, D. L. (2007) *Comparison of common root cause analysis tools and methods, in Apollo Root Cause Analysis – A New Way of Thinking*, http://kscddms.ksc.nasa.gov/Reliability/Documents/Ganos_comparison_the_Root_Cause_Analysis_Methods.pdf (accessed January 31, 2016).

Case 7

Efficiency and Air Emissions

Your position with a large US oil refiner has brought you to a small development team with the task of finding a way to improve a refinery's energy efficiency and lower its carbon dioxide (CO_2) emissions. The team includes a refinery representative, a business developer, a business analyst, and an environmental specialist. Your position on the team represents engineering. The team's recommendations will be presented to the refinery management. If successful, the team will take the refinery's concurrence to the corporation's senior management.

A key element of the analysis is related to environmental air emissions. The state's environmental laws are more progressive than federal regulations, and include a new tax on carbon emissions. After a series of brainstorming sessions the team, aware of the new regulations, has developed a list of potential projects that could satisfy the team's charter. On the list, the highest ranked project in terms of efficiency and emissions is to replace the refinery's two aging steam boilers, which provide steam for essential operations with a cogeneration facility.

Preliminary economics indicate that a nominal 90 MW gas turbine plant with heat recovery for steam production could be a successful alternative to the steam boilers and the purchase of electricity. In addition to an efficiency gain, CO_2 attributed to the facility would be reduced; therefore the taxes paid by the refiner would also decrease.

The specific plant your team has considered is a 2 × 2 combined heat and power (CHP) plant using two GE LM6000™ PC Sprint® gas turbines and two associated heat recovery steam generators (HRSGs), each having supplemental duct firing. The CHP (also referred to as cogeneration) plant would provide 95.5 MW of output, 10.5 MW of which would be sold to the local electric utility. Emissions of CO_2 attributed to the refinery would offset taxes passed on to the refinery from the utility for electricity purchases. The refinery fuel currently consumed by the boilers would be blended with natural gas to provide the fuel for the two gas turbines. Refinery electric power purchases, after construction of the CHP, would largely

Case Studies in Mechanical Engineering: Decision Making, Thermodynamics, Fluid Mechanics and Heat Transfer,
First Edition. Stuart Sabol.
© 2016 John Wiley & Sons, Ltd. Published 2016 by John Wiley & Sons, Ltd.
Companion website: www.wiley.com/go/sabol/mechanical

be eliminated with the exception of maintenance and forced outages at the new CHP plant. Slight duct firing with natural gas in the HRSGs (136.2 MMBtu/h – average) would provide enough steam-generation flexibility to follow the refinery's variable steam demand.

The refinery has had a long, respectful relationship with the local environmental authorities governing the permits for the refinery operations and new projects. In preliminary discussions, the regulators have supported the use of cogeneration because of its high energy efficiency and environmental performance. They have also made it clear that reductions in sulfur emissions compared to the existing boilers would be required to secure an operating permit for a new cogeneration facility. The maximum allowed one-hour emission rate of sulfur for the CHP, expressed as SO_2, would need to be less than 1 ppm by volume on a dry basis (ppmvd) at 15% oxygen in the exhaust gas. Competitors of your company have installed cogeneration plants within their facilities that comply with the emission rate; therefore the regulators view the requirement as reasonable and non-negotiable. Any SO_2 emission would carry a charge of $1500 per ton per annum.

Through discussions with GE, you are confident that the gas turbines can consume the gaseous fuel provided by the host refinery. General Electric has provided the data in Table 7.4 for the LM6000™-PC Sprint®[1] at the average ambient temperature at this site (59 °F) with natural gas. For preliminary calculations, the performance on natural gas is assumed to be reasonably accurate for the site fuel mixture.

This case study has a fairly typical mix of units for US facilities. Over the past 100 years, changes in conventions and global cooperation on a range of issues has led to a blending of SI and customary units at existing facilities. The author leaves it up to the student to determine which system is more convenient to address the objectives of the case study.

7.1 Background

7.1.1 Cogeneration or CHP

Cogeneration, also known as combined heat and power (CHP), provides electric power and steam for process heat to one or more host facilities, often chemical plants or refineries. A conceptual diagram of a CHP facility is shown in Figure 7.1. In other examples, steam or hot water from a CHP plant could be used for district heating systems, or as the heat source for desalination. Excess power generation from the CHP can be sold as merchant power to the grid, through a long-term power purchase agreement to a third party, or both. By providing power and steam, CHP plants have a higher energy efficiency for a host facility compared to purchasing power from a utility, and generation steam in boilers. The efficiency improvement over pure power production is derived from capturing the latent heat in steam produced at the power plant as usable energy for the hose facility; energy that would otherwise be transferred to the environment as waste heat from a steam turbine.

Combined heat and power plants collocated with a host facility can benefit by sharing host services including operations and maintenance, utilities such as water treatment, use of waste water, and very often the consumptions of a byproduct fuel. The fuel may be a gas or a liquid, but is most often a gas from a refinery or chemical plant. The composition of the fuel gas depends on the feedstock and chemical processes within the facility, and very often contains

---

[1] LM6000™ is a trademark pending and Sprint® is a registered trademark of the General Electric Company.

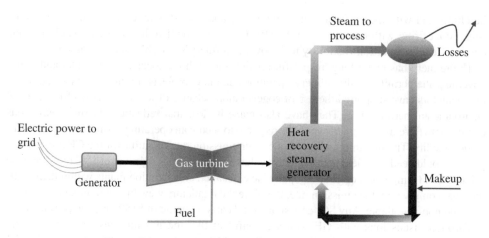

Figure 7.1 Schematic of gas turbine powered cogeneration.

high levels of hydrogen, varying amounts of olefins, and sulfur-containing compounds. Therefore, the CHP and the host facility share the benefits of cogeneration.

In a refinery, steam is an essential energy source driving the distillation processes. Any upset in the delivery of steam could result in a costly refinery outage, or failure to meet product quality. Therefore, cogeneration facilities are constructed with layers of steam reliability and must operate at a power output level necessary to produce the required steam at all times.

7.1.2 Environmental Considerations

Emissions of SO_X and NO_X from combustion sources may be regulated in multiple ways, including:

- The concentration of the emission in the exhaust gas at the emission point.
- Total mass of emissions per year.
- The maximum one-hour emission rate.

To maintain air-quality standards, an operating permit may have emission limits in all categories. SO_X and NO_X emission quantities may be expressed in ton/yr (short tons per year) in the United States, or te/yr (tonne per annum) elsewhere.

Concentrations of SO_X and NO_X in exhaust gases are expressed in parts per million on a volume basis for moisture free gas at a reference oxygen concentration. For gas turbines, the reference oxygen concentration is 15% and for boilers 3%, worldwide. Conversion of the actual concentration to that at the reference oxygen percentage is found through equation (7.1):

$$F_r = \frac{0.2095 - O_2\left(reference\right)}{0.2095 - O_2\left(dry\right)} \tag{7.1}$$

where:

O_2(reference) = fraction, 0.15 for gas turbine power plants and 0.03 for boilers;
O_2(dry) = actual O_2 volume concentration in the dry exhaust gas, fraction.

The factor, Fr, can be used to calculate the total moles of oxygen in the dry exhaust gas at the reference oxygen contraction using equation (7.2). Equation (7.2) requires knowing the total moles of dry exhaust gas and moles of oxygen in the exhaust after combustion:

$$Reference\ mole\ flow\ rate\ O_2$$
$$= \left(\frac{1}{F_r - 1}\right)\frac{(total\ dry\ exhaust\ mole\ flow\ rate)}{4.7738} \tag{7.2}$$
$$+\ mole\ flow\ rate\ O_2\ following\ combustion$$

The flow rate, in moles, of the remaining constituents of air present in the dry exhaust gas at the reference oxygen concentration can be calculated with equation (7.3):

$$moles_i = \Delta O_2 \beta_i + moles_i\ (after\ combustion) \tag{7.3}$$

where:

ΔO_2 = reference moles O_2 – moles O_2 after combustion
β_i = volume ratio of component i to oxygen in dry air
i = Ar, N_2, and CO_2

For this case study, assume that NO_X is produced at a rate of 0.198 kg per MW of fuel burned (0.123 lb_m/MMBtu) in a steam boiler burning refinery off gas (ROG) and 0.144 kg per MW of fuel consumed (0.093 lb_m/MMBtu) in an HRSG burning natural gas.

Emissions of CO_2 are reported in metric tonnes (1000 kg = 1 te) per annum throughout the world. Emissions of CO_2 from purchased power are attributed to a refinery or chemical processing facility as indirect emissions, as if the facility generated the emissions internally. The CO_2 emissions can be determined with factors and procedures that are available from the American Petroleum Institute (American Petroleum Institute, 2004). They may also be obtained directly from the US Department of Energy (Department of Energy, 1994).

In cases where a cogeneration facility provides steam and power to a host facility, the host often desires to show the emissions attributed to each utility – steam and power – separately. For a cogeneration facility, an assumption of the efficiency to produce steam is required to complete the assessment. For this case study, follow the UK procedure found in the API compendium mentioned earlier (American Petroleum Institute, 2004). The equations used for the UK procedure are equations (7.4) and (7.5), presented here for reference:

$$CO_2\ Electricity\left(\frac{te}{MWh}\right)$$
$$= \frac{2 \times CO_2\ direct\ emissions\,(te)}{2 \times electricity\,(MWh) + steam\ produced\,(MWh)} \tag{7.4}$$

Table 7.1 Fuel emissions rates and weighting factors.

	Coal (te/MWh)	Natural gas combined cycle (te/MWh)	Weighting factor
CO_2	0.894	0.432	1
Methane	1.81E-5	6.8E-6	23
N_2O	1.54E-4	2.86E-5	293

Source: API (2004).

$$CO_2 \, Steam\left(\frac{te}{MWh}\right) = \frac{CO_2 \, direct \, emissions\,(te)}{2 \times electricity\,(MWh) + steam \, produced\,(MWh)} \qquad (7.5)$$

In addition to CO_2 produced from the combustion of fossil fuels, volatile organic compounds (VOCs) released from the fuel, and N_2O produced from combustion must be accounted as equivalent CO_2 emissions. Students may use the emission rates and weighting factors in Table 7.1, or obtain current rates and rating factors from the API compendium referenced above or another credible source. The weighting factors show the greenhouse gas equivalent compared to CO_2.

For new CHP installations of the size contemplated by the development team, a selective catalytic reduction (SCR) system would be required in the HRSG to reduce NO_X emissions. An SCR catalyzes the reaction of NO_X with ammonia to produce nitrogen and water as shown in equation (7.6):

$$2NO_2 + 4NH_3 + O_2 \rightarrow 3N_2 + 6H_2O \qquad (7.6)$$

As a starting point for emission rates and eventually capital cost estimates, the team has assumed that the SCR will reduce NOx emissions by 80%. Further discussions with the regulators would determine the actual requirements for the project and the consequential capital cost for the SCR.

7.1.3 Efficiency

The useful energy in the steam and electricity divided by the fuel chemical energy required to produce the two products represents the fuel efficiency for the utilities provided to the refinery. The useful energy in the steam is taken as the difference in enthalpy between the steam delivered and the saturated liquid enthalpy at the site reference temperature. Though the actual process cannot return steam condensate at the ambient temperature, the useful energy, and therefore the value of the steam for economic and efficiency calculations, is established at the local average ambient temperature.

There are various ways to determine the value of steam for a CHP project. The combination of the selected unit energy value and the volume of steam are two of the components in the negotiation of the price of steam bought by the host and sold by the CHP. The unit energy value presented in this case study is an example method that may be used.

Table 7.2 Service territory average generation mix.

Source	Percentage	Heat rate (Btu/kWh) HHV	Heat rate (kJ/kWh) HHV
Coal	45.31	10 600	11 184
Natural gas	23.61	7475	7887
Nuclear	20.48	9724	10 259
Hydro	6.96	0	0
Renewables	3.63	0	0
Total	100	8560	9031

Source: Energy Information Agency.

Fuel required to produce electricity purchased from a utility, and therefore the indirect CO_2 emission for purchased power, is difficult to establish. The actual quantity of fuel per MWh may change in real time depending on the utility's system load and the units in service at the time. Recognizing this difficulty, the fuel required may be approximated with a blended average heat rate for the utility system for economic calculations. Published records for the local utility serving the refinery show the percent of annual generation from various sources with their average heat rate. These are reproduced in Table 7.2. The useful energy in the electricity is the full value of the power purchased in MWh.

Fuel required by a steam boiler may be calculated by an energy balance. A gas turbine's fuel requirement is provided by the manufacturer, usually in terms of the LHV of the fuel. For comparison with steam, and for economic calculations, the manufacturer's LHV fuel requirement must be converted to HHV. The concept of LHV is irrelevant for steam, and the unit price for gas fuels is generally based on higher heating value around the globe.

7.2 Case Study Details

7.2.1 General

The refinery in this case study first went into operation prior to 1940, before the adoption of SI units. It is located in the United States where there is an average ambient temperature of 59 °F. On average, 85 MW of electric power is required for the refinery operations, which is currently purchased from the local utility. The utility has provided reliable power over the years and it is willing to provide backup power in the event the refinery installs a cogeneration plant.

In addition to steam generated by the refinery processes, it requires 194 000 lb/hour of high-pressure (HP) steam at 660 psia and 570 °F, and 213 000 lb/h of medium-pressure (MP) steam at 195 psia and 470 °F. The steam must be very reliable, thus the development team anticipates keeping one of the two existing boilers available for backup during maintenance and unexpected outages at the proposed cogeneration plant. This concept would represent an improvement in reliability, which may help with acceptance of the project by the refinery management.

7.2.2   Proposed CHP Plant

A thermodynamic heat balance of the proposed CHP plant is shown in Figure 7.2. The figure shows details of a single gas turbine and the performance of two HRSGs generating steam with the exhaust from two gas turbines. As shown, the two gas turbines produce just over 95.5 MW, after netting auxiliary power requirements to run the cycle and transformer losses. Fuel to the gas turbine is compressed from normal pipeline pressure for injection into the gas turbine combustors.

The HRSG is comprised of three pressure levels, each with a steam drum. The low-pressure (LP) steam drum provides steam for the deaerating feedwater heater. Such a direct contact heater would typically be provided as an integral part of the HRSG.

The intermediate pressure (IP) drum provides steam to a tube bundle section that provides the necessary temperature for the MP process steam to the refinery. The HP drum, likewise, provides steam to a superheating tube bundle that heats the steam to the required temperature for the HP steam to the refinery. Duct firing with natural gas, located in the gas stream downstream of the HP superheater section, provides additional energy to the HRSG for steam generation.

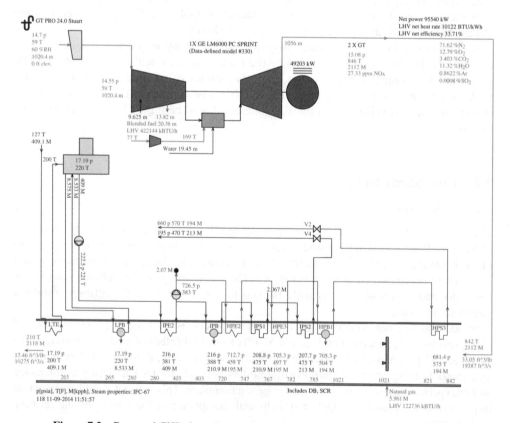

Figure 7.2   Proposed CHP plant. *Source:* Reproduced by permission of ThermoFlow.

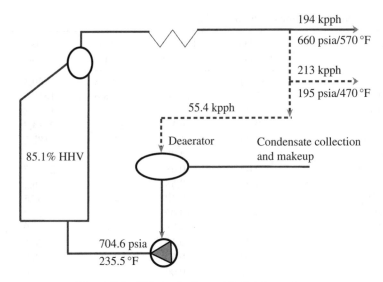

Figure 7.3   Existing boiler and feedwater system.

7.2.3   Steam Boilers

The current refinery steam generating equipment consists of two identical steam boilers that have an efficiency of 85.1% based on the HHV of the fuel. The fuel to the boilers is comprised of a mixture of refinery off gas (ROG) and natural gas, which makes up the balance of the fuel not available from ROG.

Figure 7.3 shows a sketch of the boiler feedwater, and steam systems. Medium-pressure steam is manufactured from a pressure letdown of HP steam that is tempered with feedwater flow to the required temperature. Within the boiler system, MP steam is used for feedwater heating in a direct contact deaerating heater. The feedwater enters the boiler at 704.6 psia and 235.5 °F. Makeup demineralized water completes the feedwater requirements for condensate or steam that is consumed by the refinery.

7.2.4   Fuel

The refinery produces a byproduct gaseous fuel (refinery off gas or ROG) that must be consumed by the steam boilers or the new cogeneration facility. The ROG has high hydrogen and sulfur concentrations compared to natural gas. Various olefins are also present making it unacceptable for blending into the local natural gas pipelines. Incinerating the gas (flaring or combustion in a thermal oxidizer) would not be permitted. The average production of ROG is 10.23 million standard cubic feet per day (scfd) and its composition is shown in Table 7.3 with that of the local pipeline natural gas.

7.2.5   Gas Turbine

The development team has screened several candidate gas turbines to replace the aging boilers and settled on the GE LM6000™-PC Sprint®. This model is an aeroderivative machine with a power turbine added to the base gas turbine, which drives an electric generator. The unit has

Table 7.3 Fuel gas composition and heating values.

Component	Natural Gas (%vol)	Ref gas (%vol)	LHV (kJ/mol)	LHV Btu/scf	HHV Btu/scf
Methane (CH_4)	95.20	48.85	802.6	909.3	1009.9
Ethane (C_2H_6)	2.50	13.20	1428.6	1618.5	1769.4
Ethene (C_2H_4)	0.00	0.87	1323	1498.8	1599.4
Propane (C_3H_8)	0.20	9.52	2043.2	2314.7	2515.8
Propene (C_3H_6)	0.00	1.75	1926.2	2182.2	2333.1
Butane (C_4H_{10})	0.06	4.28	2567.3	2908.6	3160.0
cis-2-butene (C_4H_8)	0.00	0.68	2533.9	2870.7	3071.9
Pentane (C_5H_{12})	0.02	0.58	3244.9	3676.2	3978.0
Hexane+ (C_6H_{14})	0.01	0.10	3855.1	4367.5	4719.5
Hydrogen (H_2)	0.00	19.05	241.8	274.0	324.3
Nitrogen (N_2)	1.29	0.78	0	0.0	0.0
Carbon monoxide (CO)	0.00	0.10	283	320.6	320.6
Carbon dioxide (CO_2)	0.70	0.19		0.0	0.0
Oxygen (O_2)	0.02	0.00		0.0	0.0
Hydrogen sulfide[a] (H_2S)	0.00	0.05[a]	518	586.9	637.1
Total	100.00	100.00			

Note: [a] Total sulfur: maximum 1 hour sulfur concentration is 800 ppm by weight expressed as H_2S.
Source: Green and Perry (2007) Reproduced with permission of McGraw Hill.

Table 7.4 Gas turbine performance data at 59 °F.

Gas turbine data (each)	
Output (kW)	49 120
Heat rate (Btu/kWh) LHV	8582
Fuel flow (lb/h)	22 188
Fuel flow (MMBtu/h) LHV	421.6
NOx water injection (lb/hr)	19 448
Sprint water flow (lb/hr)	9599
Exhaust flow (lb/s)	293
Exhaust temperature (°F)	847
NOx emissions (lb/h)	43
Net plant output (MW)	94.6

Notes: Natural gas fuel. Sea level at 59 °F ambient temperature with 60% relative humidity, 4" inlet pressure drop, 10" exhaust pressure drop, 13.8 kV generator and 0.85 power factor.
Source: General Electric.

a Sprint® package that increases power output by spraying water into a midpoint of the compressor. The water spray cools the compressed air, which reduces compressor power and adds mass flow for additional output. Water is injected into the combustion section to reduce the production of NO_X. General Electric has provided the performance listed in Table 7.4 for natural gas.

Table 7.5 Ambient air composition.

Element	Vol (%)
Ar	0.92
N_2	77.30
O_2	20.74
CO_2	0.03
H_2O	1.01
Total	100

Source: NASA.

For the case study, the gas turbine will consume all the ROG and supplement the fuel supply with natural gas. Gas-turbine exhaust, together with supplemental duct firing in the heat recovery steam generator (HRSG), will produce the refinery's steam requirement and provide flexibility to follow the variable steam demand.

7.2.6 Air

For this case study, atmospheric air at 60% relative humidity has the composition shown in Table 7.5.

7.3 Refresher

7.3.1 Gas Mixture Molecular Weight

The molecular weight of a gas mixture is the sum of the products of the constituent molecular weights and the constituent volume fraction. In Excel, use the SUMPRODUCT function as in the example shown in Table 7.6 to calculate the molecular weight.

7.3.2 Gas Mixture Heating Value

Gas heating value is calculated in a manner similar to the mixture's molecular weight. In Excel, use the SUMPRODUCT function to multiply the volume fraction of each constituent by the heating value expressed in chemical energy per unit volume of that constituent.

7.3.3 Species Weight Fraction

The weight fraction of a constituent of a gas mixture is the gas volume fraction multiplied by its molecular weight and divided by the gas mixture molecular weight as shown in equation (7.7).

$$weight\ fraction = \frac{\dfrac{mol_i}{mol_{mixture}} \cdot \dfrac{mass_i}{mol_i}}{\dfrac{mass_{mixture}}{mol_{mixture}}} \tag{7.7}$$

Table 7.6   Gas mixture molecular weight.

	A	B	C
	Fuel property data		Natural gas
		mol. Wt.	**(%vol)**
7	Methane	16.043	95.20
8	Ethane	30.069	2.50
9	Ethene	28.054	0.00
10	Propane	44.096	0.20
11	Propene	42.080	0.00
12	Butane	58.123	0.06
13	cis-2-butene	56.107	0.00
14	Pentane	72.150	0.02
15	Hexane+	86.177	0.01
16	Hydrogen	2.016	0.00
17	Nitrogen	28.013	1.29
18	Carbon monoxide	28.010	0.00
19	Carbon dioxide	44.010	0.70
20	Oxygen	31.999	0.02
21	Hydrogen sulfide	34.076	0.0005
	Total		**100.00**
	MW		16.85

MW = SUMPRODUCT($B$7:$B$21,C7:C21)/100.

7.3.4   Ultimate Analysis

The ultimate analysis of a fuel is the weight fraction of each atom making up the fuel. To find the ultimate analysis of a gas, multiply the number of each atom of each gas constituent by the atom's molecular weight and the volume fraction of the constituent. Sum the above products for all constituents and divide each sum by the mixture molecular weight. As an example, the weight fractions of carbon and hydrogen in a pure methane fuel would be:

$$\text{Carbon weight percentage: } 12.011 \cdot \frac{1}{16.043} = 74.87\%$$

$$\text{Hydrogen weight percentage: } 1.0079 \cdot \frac{4}{16.043} = 25.13\%$$

7.4   Objective

To be successful, the development team needs to be able to present a convincing case to the refinery management that the capital cost is justified by the earnings potential, and fuel savings of the cogeneration facility. The team must also present the environmental emissions profile to management in a way that would support the project. The team has chosen to show comparisons between (i) steam generated in the boilers with power purchases, and (ii) a new gas-turbine cogeneration plant. The comparisons will include:

- thermal efficiency;
- CO_2 emissions;
- compliance with SO_2 emission requirements;
- compliance with NO_x emission requirements;
- economics.

Prior to the economic evaluation, the team needs to be satisfied that:

- improvements in efficiency and CO_2 emissions can be clearly established; and
- compliance with environmental emissions limits will be met by the new cogeneration facility.

7.5 Exercises

Complete the following exercises and calculations for the existing boilers and new cogeneration facility.

7.5.1 Outside Reading

Read the sections of American Petroleum Institute (2004) on global warming potential and emissions associated with purchased electricity and answer the following questions.

1. Geothermal power production is considered a renewable source; therefore, there are no CO_2 emissions from geothermal power generation. True or False
2. One tonne of methane emission is equivalent to ___ tonnes of CO2 emission:
   (a) 23
   (b) 293
   (c) 30

7.5.2 Boiler Operation

1. Determine the composition of dry air and volume ratio of Ar, N_2, and CO_2 (β_i) to that of O_2.
2. Determine the fuel energy requirement to produce the HP and MP steam provided to the refinery.
3. Calculate the lower and higher heating values of ROG and natural gas on a volume and mass basis.
4. Determine the fuel mixture of ROG and natural gas burned in the steam boilers to satisfy the fuel energy requirement from 2. above.
5. Calculate the ultimate analysis of the fuel mixture from 4. above
6. Determine the annual average net production of equivalent CO_2, annual SO_2 and NO_2 produced in the boilers.

7. Using data in Tables 7.1 and 7.2, calculate the total CO_2 emissions attributed to the refinery from the purchase of 85 MW of power throughout the year. Students may substitute data in Tables 7.1 and 7.2 with the actual service territory generation mix and current emission and weighting values from API or another credible source.

8. Determine the fuel energy consumed each hour to produce the electricity purchased by the refinery.

9. Determine the energy value of the HP and MP steam provided to the refinery using the average ambient temperature for the reference state.

10. Calculate the energy efficiency for providing steam and power to the refinery.

11. Using 6% excess air used for combustion, calculate the concentration of SO_X, and NO_X (ppmvd) in the steam boiler exhaust corrected to 3% oxygen.

12. Determine the hourly CO_2 emissions attributed to the refinery for steam and electricity use.

13. Determine the one-hour maximum SO_2 emission concentration in ppmvd at 3% O_2.

7.5.3 Cogeneration Plant

1. Determine the gas fuel mixture molecular weight, lower and higher heating values, and the ultimate analysis for the gas turbine.

2. Determine the natural gas ultimate analysis for the HRSG duct firing.

3. Determine the air mass flow to the gas turbine

4. Determine the composition of air and gases at the following locations of the CHP plant:
   (a) Gas turbine compressor discharge.
   (b) Gas turbine exhaust.
   (c) HRSG exhaust.

5. Determine the dry exhaust gas composition at the reference oxygen level for the HRSG.

6. Determine the energy efficiency of steam and power production for the refinery.

7. Determine the annual emission rates of equivalent CO_2, NO_X and SO_X at the HRSG exhaust and the concentration of NO_X and SO_X (ppmvd at the reference O_2 concentration)

8. Determine the annual CO_2 emissions attributed to the refinery for steam and electricity production for the CHP option.

9. Determine the maximum SO_2 concentration in ppmvd at 15% O_2 for the CHP option.

7.5.4 Conclusion

1. Tabulate the key findings from the exercises above.

2. State reasons to continue or discontinue the development effort.

3. Provide a recommendation on how to advance the development of a new cogeneration facility for the refinery. Support your recommendation.

7.6 Closure

The cogeneration development option provided several benefits for various stakeholders. The energy efficiency improvement represented a cost saving to the refinery as did the tax reduction due to the 24% reduction in total CO_2 emissions attributed to the refinery. In addition to the CO_2 emission reduction, environmental performance of the refinery would be improved by lower NO_x emissions of almost 200 t/yr from direct and indirect emissions. Therefore, residence in the service area would benefit from improved air quality, shareholders of investments in the refinery would benefit by greater returns, and the corporation would benefit due to improved returns on its investments and its standing in the community.

However, the non-negotiable position related to sulfur emissions placed an economic burden on the project that was too high to overcome. The development team attempted to find a capital solution to remove the sulfur from the refinery fuel gas but the benefits could not overcome the cost. Therefore, rather than a win-win situation for the refinery and the community, the development was terminated and the refinery continued producing steam as it had.

Each party in the development process, the engineers, managers, business developers, and environmental regulators performed their duties properly and as expected. In the long term, the refinery may be forced to replace the boilers for other reasons, and the environmental objectives would be satisfied. However, the environment cost and impact on the community of delaying the conversion may offset the eventual improvement.

7.7 Symbols and Abbreviations

MW: molecular weight M/mol; megawatts
M: mass
scfm: standard cubic feet per minute
standard conditions: English: 14.696 psia, 60°F; SI 101.3 kPa, 273.15 K
βi = volume ratio of component i to that of oxygen in dry air in equation (7.3).

7.8 Answer Key

7.5.1
1. False. Steam from geothermal reserves contains CO^2, which must be vented from the steam-turbine condenser.
2. a.

7.5.2
1. Composition of dry air:

	Composition of dry air		
	vol.%	wt%	β_i
Ar	0.934	1.288	0.045
N_2	78.087	75.522	3.728
O_2	20.948	23.142	1.000
CO_2	0.031	0.048	0.001
Water	0.000	0.000	0.000
Total	100.00	100.00	
MW	28.965		

2. 571.9 MMBtu/h, 167.6 MW.

3.

	Customary (Btu/scf)	SI (kJ/scm)*
LHV ROG	1156	45 527
HHV ROG	1272	
LHV natural gas	914	35 988
HHV natural gas	1014	

*Usually standard conditions are taken as 0 °C and 101 kPa when using SI units. For this case study, 60 °F and 14.696 psia are used. No corrections have been applied for the heating value between 60 °F and 0 °C.

4.

	Customary (Btu/scf)	SI (kJ/Mm³)*
LHV mixture	1140	44 910
HHV mixture	1255	

* See notes above.

5. Mixture ultimate analysis.

	C	H	N	O	S
Wt% in mixture	76.35%	22.06%	1.09%	0.42%	0.08%

6. Equivalent CO_2: .278,425 te/y for steam; SO_2: 151 t/yr for steam.
7. Equivalent CO2: 397,151 te/y for electricity.
8. 727.6 MMBtu/h, 213.1 MW.
9. 501.2 MMBtu/h, 146.9 MW.
10. 60.91%.
11. SOX: 39.8 ppmvd @ 3% O_2, NOX: 95 ppmvd at 3% O_2, 286 t/yr.
12. Steam: 31.8 te/h, electricity: 45.3 te/h.
13. 63.6 ppmvd @ 3% O_2 at 800 ppm sulfur and full load on the boiler.

7.5.3

1. Fuel gas mixture for the gas turbine. MW = 19.08:

	Customary (Btu/scf)	SI (kJ/Nm³)*
LHV mixture	1041	41 009
HHV mixture	1150	

Ultimate analysis:

	C	H	N	O	S
Wt% in mixture	74.92%	22.74%	1.50%	0.79%	0.05%

2. Natural gas ultimate analysis:

	C	H	N	O	S
Wt% in mixture	72.65%	23.83%	2.15%	1.37%	0.001%

3. 2 007 130 lb/h, 252 894 kg/s
4. Composition in volume percentage:

	Compressor discharge (vol.%)	Gas turbine exhaust (vol.%)	HRSG exhaust (vol.%)
Ar	0.91	0.86	0.85
N_2	76.13	71.60	71.25
O_2	20.42	12.79	11.78
CO_2	0.03	3.41	3.87
H_2O	2.50	11.34	12.24
NO_X as NO_2	0.00	0.002	0.001
SO_X as SO_2	0.00	0.001	0.001

5. Dry composition at HRSG exit:

	HRSG exhaust – dry (vol.%)
Ar	0.97
N_2	81.19
O_2	13.43
CO_2	4.41
H_2O	0.00
NO_X as NO_2	0.001
SO_X as SO_2	0.001

6. 77.51%
7.

	HRSG discharge
Equiv. CO_2 (te/year)	513 531
NO_X as NO_2 (t/year)	86
SO_X as SO_2 (t/year)	168
NO_X (ppmvd @ 15% O_2)	5.1
SO_X (ppmvd @ 15% O_2)	7.1

8. Equivalent CO_2 for steam: 220 450 te/y. Equivalent CO_2 for electricity: 293 021 te/y.
9. 10.8 ppmvd @ 15% O_2.

7.5.4

1. Key findings:

	Existing refinery	Cogeneration option
CO_2 for steam (te/year)	278425	220476
CO_2 for electricity (te/year)	397151	293055
Total (te/year)	675576	513531
Energy efficiency	60.91%	77.51%
Maximum* SO_x emissions (ppmvd)	63.6 @ 3% O_2	11.4 @ 15% O_2
SO_x emissions (t/year)	151	168
NO_x emissions (ppmvd)	95.2 @ 3% O_2	5.1 @ 15% O_2
NO_x emissions (t/year)	286	86

* at 800 ppm sulfur.

2. The cogeneration option provides significant improvements over the existing refinery in terms of energy efficiency and environmental emissions of CO_2 and NO_x. However, the cogeneration option does not improve the SO_x emission rate per annum and cannot meet the regulator's concentration requirement of 1 ppmvd at 15% O_2. The improvements in efficiency and environmental performance encourage continuing the project. Failure to meet the sulfur emission requirement is a good reason to halt further development.

3. To satisfy corporate and refinery management of the correct decision on whether to halt or proceed, the development team must determine a budgetary estimate to remove the sulfur from the refinery off gas. Once determined, an economic evaluation would support the decision. If the CHP choice is to terminate, the next best alternative should be evaluated by the task team.

References

American Petroleum Institute (2004) *Compendium of Greenhouse Gas Emissions Methodologies for the Oil and Natural Gas Industry*, API, Washington DC.

Department of Energy (1994) *Sector-Specific Issues and Reporting Methodologies Supporting the General Guidelines for the Voluntary Reporting of Greenhouse Gases under Section 1605(b) of the Energy Policy Act of 1992*, vol. 1. DOE/PO-0028. DOE, Washington DC.

Green, D. W. and Perry, R. H. (2007) *Perry's Chemical Engineers' Handbook*. 8th edn. McGraw Hill, New York, NY.

Case 8

Low-Carbon Power Production*

The Sierra Club's web site claims that wind energy plays an important part in the battle against climate change and in moving away from a dependence on fossil fuels (http://content.sierraclub.org/coal/wind/wind, accessed January 27, 2016). The Wind Energy Foundation web site states that wind energy is one of the most cost-effective sources of electricity because the fuel is free and there are virtually no polluting side effects (http://www.windenergyfoundation.org/about-wind-energy, accessed January 27, 2016). The US Department of Energy states that the winds in the United States and its coastal waters can provide about four times the electric generating capacity of the 2013 US power grid (http://energy.gov/articles/top-10-things-you-didnt-know-about-wind-power, accessed January 27, 2016). The Energy.gov web site also states that wind energy could provide abundant, clean, renewable energy for the United States. Statements on these and other web sites speak to the economic benefits of wind generation, its affordability, and abundance.

In addition to the statements above in favor of wind generation, the US Department of Energy, on the Energy.gov web site, recognizes that support for wind generation, in particular the production tax credit, is essential to meet development goals for such generation. This statement implies that wind generation is not as economical as its proponents contend. It presents another side of the wind energy discussion for an investor – whether it is prudent to spend capital in an industry that requires government support to be economic. It raises questions about the future of the technology, as well as whether there are other, more economic choices that would meet societal goals, how those goals are set, and who sets them.

---

*Data presented in this case study was current at the time of writing. Students may wish to obtain data from credible sources for a specific site, or facility, or data, that is more recent.

Case Studies in Mechanical Engineering: Decision Making, Thermodynamics, Fluid Mechanics and Heat Transfer,
First Edition. Stuart Sabol.
© 2016 John Wiley & Sons, Ltd. Published 2016 by John Wiley & Sons, Ltd.
Companion website: www.wiley.com/go/sabol/mechanical

Understanding the issues, costs, and benefits in a political arena such as the discussion of wind energy can be complex. It may be difficult to sort out the differences between the facts and the facts as presented by an advocate or opponent. Whether opinions have worked to filter the interpretation of facts can be an equally important question when discussing research findings on this topic.

Engineers are often asked to develop and produce documentation for a corporate position as employees or consultants. The subject of this case study is such an example. How you respond can be just as important as the answers provided. Taking a position on the subject may work against your credibility. Therefore, independent, qualified sources, and sound research with an unbiased delivery are required.

Your company is contemplating a product line of reduction gears for the wind industry. This would not be a new area of expertise for the company but an entry into a new market. You and the company's economist are putting together a model to help analyze the industry, and you are asked for a comparative view of the economics, and carbon dioxide emissions of several power-generation technologies. The sources must be from outside your company as internal sources could appear to be biased in favor of it. The chief economist desires to present a completely independent position to the corporate executives.

Electricity is a large section of the world economy – an essential element required for the health and safety of the population, and a necessity for the production of raw materials, and finished consumer products. Power production is also a significant stationary source of priority environmental pollutants including greenhouse gases (GHG). The mining and use of natural resources required for the generation of electricity affects ecosystems, and often destroys the habitat of endangered species. Wind energy can impact the flight path of migratory birds and birds of prey. It is clear that the subject of power generation is enormous and can easily overwhelm your time and resources. You have therefore decided to narrow the focus of study.

There are several possible areas to start – electric power distribution grid stability, wind patterns, onshore and offshore wind projects, economics, political pressures, impacts on avian wildlife – to name a few. As a mechanical engineer you have decided to avoid questions on the electrical grid, weather, wildlife, and politics. Economics and emission characteristics will be your focus.

A key part of your analysis must account for the objectives of using wind generation. Gleaning ideas from the Sierra Club and the Wind Energy Foundation, you find that wind energy is part of a strategy to reduce the use of fossil fuels, thereby reducing carbon dioxide emissions and combating the effects of man-made climate change. You will therefore calculate the cost of avoided carbon dioxide emissions for various technologies compared to advanced technology coal-fired power plants.

8.1 Background

Various electric-generating technologies are often compared using the levelized cost of electricity (LCOE), or levelized cost of energy – the ratio of the lifetime cost of building and operating the facility to the lifetime energy production. The calculation is not intended to determine the actual price of electricity as there are many factors that are not included in the calculation but it can be used to compare technologies and serve as an indication of how the market will react to various pressures or incentives.

When determining LCOE for a technology it is important to state clearly the assumptions for the calculation and to use a consistent set of assumptions for the various technologies.

Different researchers or organizations may apply a variety of assumptions, which are not easily determined from the results. The author recommends performing the calculations independently to ensure a full understanding of the results.

The LCOE is a calculation of the net present value of a series of costs and earnings over a given economic life for an installation. These include the initial development costs, purchase of property, obtaining permits, the capital cost for the facility, fuel, and direct operations and maintenance required to produce electricity. Tax benefits, government incentives, and real-time dispatch decisions are generally not included in the LCOE calculation. The LCOE does not include the borrowing cost, depreciation, interest expenses or taxes on property or income that are included in calculations for an investment decision. The LCOE shows the cost to produce the power rather than the cost of power required to yield a profit for the facility owner.

From its definition, a simple LCOE formula can be developed as shown in equation (8.1). In this equation, the capital recovery factor is determined such that the net present value of capital expenditures for construction, operating cost and revenues equals zero for the life of the project (http://www.nrel.gov/analysis/tech_lcoe_documentation.html, accessed January 27, 2016):

$$LCOE = \frac{\sum_{t=1}^{n} \left\{ \frac{CR_t + O\&M_t + FC_t}{(1+r)^t} \right\}}{\sum_{t=1}^{n} \left\{ \frac{MWh_t}{(1+r)^t} \right\}} \tag{8.1}$$

where:
  CR_t = capital recovery in year t;
  $O\&M_t$ = Fixed and variable operating and maintenance costs in year t;
  FC_t = fuel costs in year t;
  MWh_t = electricity produced in year t;
  r = discount rate;
  n = economic life in years;
  t = years from 1 to n.

8.1.1   Dispatch and Renewable Power Resources

An important assumption in this analysis is the method of equating technologies capable of economic dispatch (fossil and nuclear) to renewable sources, which are dependent on weather, or the availability of sunshine. Onshore wind generation typically has an operating capacity factor (CF) of less than 35% due to weather (http://www.eia.gov/electricity/monthly/epm_table_grapher.cfm?t=epmt_6_07_b, accessed January 27, 2016). That is, something less than 35% of the installed wind capacity operates over the course of a year, and it is only available when the wind blows – mostly at night. Therefore, increasing the installed capacity of wind generators would not necessarily improve the capacity available to generate when power is most in demand.

In practice, the unreliable nature of wind generation is compensated by fossil-fired technologies. To compare wind generation to fossil and nuclear sources on the same economic and environmental basis, this case study assumes that natural gas-fired backup generation is installed to compensate for the low availability factor of wind. Specifically, natural gas generation comprised of 50% single-cycle gas turbines and 50% natural gas-fired combined cycle gas turbines (CCGTs).

8.1.2   Capacity Factor and Availability Factor

Another important assumption is the capacity factor (CF) of fossil-fired generation, which is sometimes confused with availability. The availability factor (AF) of a power generator is the percentage of time the generator is capable of providing 100% of its capacity. The capacity factor is the percentage of the installed capacity that is actually utilized over a period of time. For example, a CCGT on economic dispatch may be available to generate but, for economic reasons, may be held in reserve until demand and price are more favorable. During the period when the CCGT facility is offline for economic reasons, the capacity factor would be zero and the availability factor would be 100%. Thus the availability factor is always higher than the capacity factor over a long period of time. For LCOE evaluations, the availability factor is most often used as the basis of comparison because economic dispatch decisions are excluded from the calculation. However, at times various references site the availability factor as the capacity factor.

The EIA lists reasonable availability factors (as capacity factor) for various technologies (Energy Information Administration (2014)). These are summarized in Table 8.1.

8.1.3   Fuel Costs (FC in Equation (8.1))

In many cases, an economist working for a corporation would provide a projection of future prices for various fuels. If an economist's opinion is not available, the corporation may commission an outside consultant to provide a pricing assessment. When neither of these options is available, an independent source may be available. The EIA provides current coal and natural gas prices (http://www.eia.gov/coal/news_markets/, http://www.eia.gov/dnav/ng/hist/rngwhhdm.htm, accessed January 27, 2015), and its publication, *Annual Energy Outlook,* includes one possible version of future prices.

Note that the EIA provides coal prices at the mine, and at a major distribution hubs for natural gas. Transportation and delivery costs must be added for a complete cost of the fuel. For

Table 8.1   Availability factors.

Technology	Availability factor (%)
Advanced pulverized coal	85
Conventional CCGT	87
Single cycle gas turbine[a]	87
Onshore wind	36
Nuclear	90
Advanced pulverized coal with carbon capture and storage[b]	85

Notes: [a]Assumed to be the same as CCGT. [b]Assumed to be the same as IGCC and advanced coal.
Source: EIA (2014).

this case study, assume that coal delivery costs are $20/ton ($18.14/te) for Powder River Basin coal and $0.20/MMBtu ($0.19/GJ) transportation costs for natural gas from the Henry Hub.

Nuclear fuel prices can be determined from the Nuclear Energy Institute as $0.764/MMBtu, ($0.724/GJ) (http://www.nei.org/Knowledge-Center/Nuclear-Statistics/Costs-Fuel,-Operation,-Waste-Disposal-Life-Cycle, accessed January 27, 2016) for a nuclear plant heat rate of 10 339 Btu/kWh (10 908 kJ/kWh).

8.1.4 Capital Cost Recovery (CR in Equation (8.1))

For LCOE, the cost of building a new facility is converted into a payment series over the economic life of the project. The net present value of the series will equal the net present value of the capital cost of the new facility plus any owner's costs to develop the project. In Excel, this can be done with the goal seek function or simply by trial and error to find the capital recovery payment in $/MWh so that the total NPV of the sum of capital costs and annual revenue from the sale of electricity equal zero.

8.1.5 Nonfuel Operations and Maintenance (M in Equation (8.1))

Operations and maintenance (O&M) costs are divided into fixed and variable categories. Fixed costs are those that are independent of production, for example fixed labor, and subcontracts including security, insurance, and land leases. Fixed O&M costs for power generation are often expressed in terms of installed capacity ($/kW/yr).

Variable O&M costs are those that are proportional to production. These nonfuel costs include the cost for cooling water, ordinary maintenance of equipment, major maintenance related to operation hours, the cost of consumable materials used in maintenance, nuclear fuel disposal costs, and so forth. These costs are expressed per unit of output – $/MWh.

When presenting LCOE, fixed and variable costs may be shown separately or together, depending on the audience. For this case study, you will need to calculate the fixed O&M costs in terms of unit output ($/MWh).

8.1.6 Regulation and Government Support

Governments throughout the world regulate and promote portions of the electric power industry in a variety of ways, one of which is through the support of a chosen technology. Support can take the form of a renewable portfolio standard (RPS), or tax credits as mentioned earlier. An RPS would dictate or encourage a certain level of generation from renewable sources, the cost of which could be recovered by the supplier through a regulated power price or tax credits such as a production tax credit, mentioned by Energy.gov. The costs would be passed on to the consumer in the various forms that may include taxes on electricity usage, sales tax, income tax, and so forth.

Such programs are often enacted to give emerging technologies a foothold in a mature market. In theory, once the emerging technology achieves a level of penetration into the market, costs have decreased to the point where it can compete with established technologies without a subsidy or tax benefit.

8.2 Refresher

8.2.1 Short-Run Marginal Cost

The short-run marginal cost is the cost required to produce the next increment of a product, in this case electricity. Included in the cost are the incremental fuel and variable expenses. Incremental fuel expenses are calculated with equation (8.2):

$$fuel\,cost\left(\frac{\$}{kWh}\right) = Heat\,Rate\left(\frac{kJ}{kWh}\right)\cdot fuel\,price\left(\frac{\$}{kJ}\right) \qquad (8.2)$$

Variable operations and maintenance costs are generally presented in units of \$/MWh for most power-generation technologies.

8.2.2 CO_2 Emissions

The determination of the quantity of carbon dioxide (CO_2) emitted per annum for a technology relies on the fuel flow from heat rate and unit capacity, the weight fraction of carbon in the fuel and the unit capacity factor. Equation (8.3) shows a typical calculation of the quantity of CO_2 emitted per year. International conventions express CO_2 in tonnes (te) (1 te = 1000 kg):

$$\frac{HR_{HHV}\cdot Output}{fuel\,HHV}\cdot C\left(\frac{44}{12}\right)\cdot 8760\cdot CF \div Conv \qquad (8.3)$$

where:
  HR_{HHV}: heat rate based on the higher heating value of fuel (kJ/kWh);
  output: capacity of the generator (kW);
  fuel HHV: fuel higher heating value (kJ/kg);
  C: weight fraction of carbon in the fuel;
  CF: capacity factor;
  Conv: conversion to tonnes = 1000.

8.2.3 Long-Run Marginal Cost

The long-run marginal cost is the cost of a new entrant to a market. For this case use the LCOE as the long-run marginal cost.

8.3 Case Study Details

For this case study, you will calculate the LCOE and carbon dioxide emissions for the following generic technologies:

- advanced pulverized coal;
- CCGT;
- single unit nuclear;
- wind with and without natural gas-fired backup generation; and
- coal with carbon capture and sequestration (CCS).

Table 8.2 Development periods.

	Early development (year)	Front-end engineering and permitting (year)	Construction (year)	Development costs ($million/year)
Advanced pulverized coal	1.5	1.5	4	6
CCGT	1	1	3	4
Nuclear	2	3	4	8
Wind with backup	1	1	3	4
Wind without backup	2		1	2
Coal with CCS	2	2	4	6

Sources: National Gas Supply Organization and Southern Company.

Each technology will carry the following assumptions:

- Economic life of 30 years.
- Discount rate, 10%.
- Unit capacity 650 MW.
- Powder River Basin coal at $11.55/ton ($10.48/te). (http://www.eia.gov/dnav/ng/hist/rngwhhdm.htm) at the mine with a heating value of 8564 Btu/lb (19,920 kJ/kg) (National Energy Technology Laboratory, 2012).
- Natural Gas at the Henry Hub: $3.86/MMBtu ($3.66/GJ).
- For inflation use a Consumer Price Index of 2.5% per year. All expenditures for fuel, operations and maintenance costs escalate with the CPI.
- Planning to build a new facility in each technology would begin immediately. Actual construction would begin after the completion of planning, preliminary and detail engineering and after the receipt of permits. The assumptions in Table 8.2 may be used for each technology (Natural Gas Supply Organization, 2015; http://www.southerncompany.com/what-doing/energy-innovation/nuclear-energy/pdfs/vogtle-nuclear-brochure.pdf, accessed January 27, 2016).

Additionally, you will calculate the carbon emissions for petroleum-based generation and wind with coal-fired backup generation.

8.3.1 Reading Assignment

Locate a source or set of sources that contain the following information for the five generating technologies listed in Section 8.3 for which LCOE will be calculated:

- capital cost: ($/kW);
- fixed and variable O&M;
- availability factor;
- heat rate (Btu/kWh or kJ/kWh);
- transmission differential ($/MWh).

Most resources provide the capital cost for a dual-unit nuclear facility much larger than the case study. The installation costs for the dual facility include efficiencies gained by building two units on a single property at nearly the same time. In order to find the construction cost in ($/kW) for a single-unit facility, use equation (8.4):

$$Single\ unit\left(\frac{\$}{kW}\right) = Dual\ unit\left(\frac{\$}{kW}\right)\cdot 2^{0.4} \tag{8.4}$$

8.3.2 Transmission Costs

Wind generation in the United States is most productive in a region stretching from North Dakota to West Texas, see Figure 8.1 (http://energy.gov/eere/wind/wind-resource-assessment-and-characterization, accessed January 28, 2016). Population centers are far from this region. To bring wind generation to the population centers, where the generation is used, requires additional transmission infrastructure and capital costs. Offshore wind generation would have a similar obstacle to overcome. For this case study, you may utilize the transmission differential between CCGT and wind or add the transmission requirement to each technology.

Develop a table by technology for a LCOE calculation showing the capital cost of construction, heat rate, availability factor, fixed and variable O&M costs, and the cost

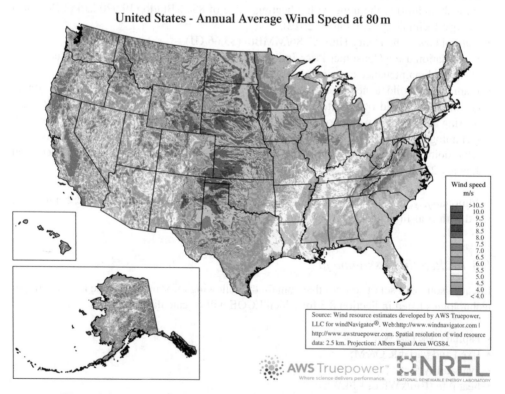

Figure 8.1   US wind resources. *Source*: National Renewable Energy Laboratory.

of transmission system upgrades. The author suggests using the Energy Information Administration as a source of information.

8.3.3   Economic Models

For each technology, develop a cash flow model for the calculation of LCOE, showing expenses and earnings on a before-tax basis. The model should include the following revenues and costs for a 30-year economic life after the completion of construction:

- development and capital cost to construct each facility;
- the capital recovery revenue necessary to recover the investment required to develop and construct new facilities;
- variable O&M;
- fixed O&M;
- fuel costs;
- transmission costs.

Summarize your findings in a stacked bar chart for each technology. Rank the technologies as a function of their LCOE.

8.3.4   Carbon Emissions

1. Calculate the carbon dioxide emissions per MWh for advanced pulverized coal, conventional natural gas CCGT and petroleum based generation. Use the following assumptions:
   - Powder River Basin coal and petroleum composition with the properties shown in Table 8.3 (Speight, 1997; Morvay and Gvozdenac, 2008: Part III; National Energy Technology Laboratory, 2012);
   - natural gas as pure methane with a higher heating value of 22 500 Btu/lb (52 335 kJ/kg);
   - petroleum fuel heat rate: 9900 Btu/kWh (23 027 kJ/kg).

Table 8.3   Powder River basis coal and petroleum composition.

	Coal	Petroleum
Carbon (C) wt. %	50.07	83–85
Hydrogen (H) wt. %	3.38	11–12
Nitrogen (N) wt. %	0.71	0–1
Sulfur (S) wt. %	0.73	2–4
Oxygen (O) wt. %	11.14	0
Moisture wt. %	25.77	0
Ash wt. %	8.19	0–0.12
Higher heating value (Btu/lb/kJ/kg)	8564/19 920	18 800–20 500/43 700–47 700

Sources: Speight (1997); Morvay and Gvozdenac (2008): Part III; National Energy Technology Laboratory (2012).

Table 8.4 System generation mix.

Technology	Generation (%)
Coal	37.4
Natural gas CCGT	30.3
Nuclear	19
Hydro	6.8
Wind	3.5
Petroleum	0.3
Other (biofuel)	2.7
Total	**100**

Source: EIA (n.d.).

2. Calculate the system's CO_2 emissions and LCOE given the generation mix of Table 8.4 (Energy Information Administration, n.d.). Use the following LCOE values
   - hydro: $99/MWh;
   - petroleum $112/MWh;
   - bio: $120/MWh.

8.3.5 Understanding the Findings

The following exercises create concrete examples of how market pressure may impact the system mix and the LCOE for the example system mentioned above:

1. Generators are dispatched in the order of increasing short-run marginal cost (the sum of fuel and variable operations and maintenance costs). Rank the following in dispatch order:
   (a) Coal.
   (b) Natural gas CCGT.
   (c) Nuclear.
   (d) Wind.
2. The governing authority of problem 8.3.4, item 2. above has an option to pass a statue that will call for the reduction of CO_2 emission. It has the option of setting a renewable portfolio standard (RPS), or requiring a certain percentage of coal generation to be replaced by either nuclear or natural gas CCGT. Assume that renewable generation displaces the generation technology with the highest short-run marginal cost from problem 8.3.3 above. For the following three options, calculate items a. through d. below. Option 1: 15% RPS. Option 2: 20% replacement of coal with nuclear. Option 3: 20% replacement of coal with natural gas CCGT.
   (a) System LCOE.
   (b) CO_2 emissions (te/MWh).
   (c) The change in LCOE from the base of problem 8.3.3.
   (d) The cost of avoided CO_2 emissions ($/te).
3. Rather than set a required makeup of system generation, the governing authority has the option to tax the emissions of CO_2.
   (a) If the tax is to be $10/te, would there be any change in the market behavior?

    (b) What is the CO_2 tax ($/te) that would result in a shift from coal to natural gas generation based on the short term marginal cost of generation.

    (c) Round up the answer from b. above to the nearest $10/te and calculate the following with the assumption that nearly all (80%) coal generation will be replaced by natural gas CCGT:

        (i) System LCOE

        (ii) CO_2 emissions (te/MWh)

        (iii) The change in LCOE from the base of problem 8.3.3.

        (iv) The cost of avoided CO_2 emissions ($/te)

4. Wind generation in the United States grew significantly between 2000 and 2014 due to government-imposed renewable portfolio standards, and benefits such as the product tax credit among other state-run programs. When the RPS goals are met, and the tax credits expire, will the robust growth of wind generation continue? Why?

5. When a web site mentions that wind power is a cost-effective source of electricity, what does the site mean? What is missing from the site's discussions?

8.3.6 Explaining the Results

It is important in these types of studies to be able to explain the results in simple straightforward terms. Understanding the results well enough to explain them in such terms often requires answer the questions "Does this make sense?" That is, can the results be inferred from the basic facts? For example, in section 8.3.5, exercise 2 imposed an RPS on the system. Construction of wind generation would displace the highest marginal cost units – CCGT; but these are low emitters of CO_2. Displacing CCGT leaves the higher emitting coal units in service, thereby reducing the impact that wind would have on the system CO_2 emission rate. Further affecting the emission result is the low capacity factor of wind generation. The implied conclusion is that reduction in CO_2 should be lower than the RPS, which is shown by the results. Alternatives that replace coal generation with generation that has little or no CO_2 emissions would have a greater impact. Therefore, understanding the short run marginal cost, and relative CO_2 emission rates helps explain the results.

    Another clear finding from these calculations is that wind generation is much more expensive to build than CCGT or hydro-power as shown by LCOE. Since hydro-electric capacity is dictated by geology and can be difficult to permit, most developers would prefer CCGT as an option well before wind in the absence of a requirement or government sponsored incentives. This result is supported by the Energy.gov website, which has stated that wind generation requires incentive support. Once the support ends, the market for new wind generation would end as well.

    The LCOE also supports the cost efficiency of a CO_2 emission tax. The tax would drive high emitters out of the market – coal also has nearly one of the highest LCOE. Therefore, economic pressure, in the form of a tax, results in less emission and lower electricity prices than any of the other alternatives examined.

8.4 Closure

The research showed that selection of new wind-generation facilities over competing technologies required government support. Given that government incentives would terminate in the near future, the company decided to postpone further development of a new product line.

A few calls to colleagues at companies developing wind-generation assets showed that those companies were scaling back their operations. Some had already experience layoffs.

The research showed that the wind industry was not resilient without government support in the form or credits, portfolio requirements, or taxes on power generation. The results showed that, compared to enacting a 15% RPS or regulating replacement of coal-fired generation with nuclear power, a tax on GHG would yield a lower long-run marginal cost of electricity, with greater reduction of GHG emissions.

Estimates showed a 29% increase in the short-run marginal cost of natural-gas-fired combined-cycle generation due to a GHG emission tax of $40/te. The chief economist felt that a short-term increase of that magnitude would be opposed at a level that would make it difficult to enact. Therefore, while making a direct connection between the cost and the target emission, and lessening emissions at the least avoided cost, it was probably not a viable alternative for the near term. Even if enacted, the LCOE for natural gas CCGT would be below that for wind. The economics would therefore not encourage additional construction of wind facilities.

In order to achieve greater reductions in GHG emissions, the percentage of wind in the generation mix would have to be large enough to replace coal generation, which had a lower short-run marginal cost than natural gas. At that point, the company felt that the electric grid stability would begin to be effected due to the variable nature of wind and the long startup times required by coal generators. Displacement of coal would also decrease generating efficiency thereby reducing the GHG benefits (Bentek Energy, LLC, 2010). There therefore appeared to be a diminishing rate of return for adding wind generation – a fact that could impact future government intervention.

Since public opinion could change, government support for wind could fluctuate. The company felt that reliance on public opinion added another layer of risk to the proposed business venture, and they turned their attention elsewhere.

8.5   Answer Key

Section 8.3.1: Data Provided by EIA (Energy Information Administration, 2013: Table 1) is shown in Table 8.5.

Table 8.5   LCOE calculation inputs.

	Advanced coal	CCGT	Onshore wind	Nuclear[a]	Simple cycle GT	Advanced coal with CCS
Capital cost ($/kW)	$3246	$917	$2213	$7297	$973	$5227
Heat rate (Btu/kWh/ kJ/kWh) HHV	8800/ 9284	7050/ 7438	NA	10339/ 10908	10850/ 11447	12000/ 12661
Availability factor (%)	85%	87%	35%	90%	87%	85%
Fixed O&M ($/kW)	$37.80	$13.17	$39.55	$93.28	$7.34	$80.53
Variable O&M ($/MWh)	$4.47	$3.60	$0.00	$2.14	$15.45	$9.51
Transmission ($/MWh)[b]	0	0	$2.40	0	0	0

Notes: [a]Dual nuclear facility scaled to single unit using 0.6 power; [b]Compared to natural gas-fired CCGT.
Source: Energy Information Administration (2013: Table 1).

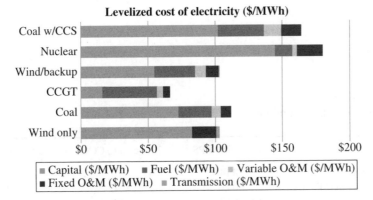

Figure 8.2   Stacked bar chart of LCOE results.

Section 8.3.3: Economic Model Results with Stacked Bar Chart Shown in Figure 8.2

	Wind only ($)	Coal ($)	CCGT ($)	Wind/ backup ($)	Nuclear ($)	Coal w/CCS ($)
Capital ($/MWh)	82.49	72.52	16.09	54.84	144.72	102.05
Fuel ($/MWh)	0.00	24.67	40.49	30.13	12.85	34.35
Variable O&M ($/MWh)	0.00	6.80	4.58	8.06	3.48	13.03
Fixed O&M ($/MWh)	18.10	7.73	5.20	9.43	19.24	14.78
Transmission ($/MWh)	2.40	0.00	0.00	0.83	0.00	0.00
Total	**102.98**	**111.72**	**66.36**	**105.30**	**180.29**	**164.21**

Section 8.3.4: Carbon Emissions

1.

	Coal	CCGT	Petroleum
CO_2 emissions (te/MWh)	0.86	0.41	0.8

2.   Table 8.6 shows the base-case CO_2 emissions and LCOE for system generation mix.

Table 8.6   System LCOE calculation.

	% Gen*	te CO_2/MWh	Portion of CO_2/MWh	LCOE ($)	Portion of total LCOE
Coal	37.4	0.86	0.32	112	42
Natural gas	30.3	0.42	0.13	66	20
Nuclear	19.0	0	0.00	180	34
Hydro	6.8	0	0.00	99	7
Renewables (wind only)	3.5	0.00	0.00	106	4
Petroleum	0.3	0.68	0.00	112	0
Other	2.7	0.00	0.00	120	3
Total	**100.0**		**0.45**		**110**

Source: EIA.

Section 8.3.5

1. Dispatch rank
   (i) Wind
   (ii) Nuclear
   (iii) Coal
   (iv) Natural gas.
2. Government regulatory options:

Option	15% wind RPS	20% coal to CCGT	20% coal to nuclear
System LCOE	114	107	115
CO_2 emissions (te/MWh)	0.40	0.42	0.38
ΔLCOE	−10.7%	−7%	−14%
Avoided CO_2 ($/MWh)	$88	−$104	$80

3. CO_2 tax
   (a) No. A $10/te emission tax on CO_2 emissions would not increase the short run marginal cost of coal over the next most expensive fuel source – CCGT. All other factors remaining equal, there would be no change in the economic dispatch of power generators.
   (b) Equity in the short run marginal cost for advanced coal fired generation and conventional natural gas fired CCGT occurs at a CO_2 emission tax of about $31/te.
   (c) At $40/te for CO_2 emissions:
       (i) System LCOE = $109/MWh (0.8% decrease from the base case)
       (ii) CO_2 emissions = 0.32 te/MWh (29% reduction from base case)
       (iii) Avoided CO_2 = -$342/te
4. After government support programs expire, there will likely be little growth in wind generation. The competing technologies have a lower LCOE than wind. In the absence of a tax on GHG gases, the costs of man-made climate change are not reflected in the cost of electricity. Behavior governed by economic decisions would result in less growth of wind generation.
5. The low cost of wind generation refers to its short-run marginal cost. Missing are the levelized capital cost of construction, fixed O&M costs, and the increased cost of transmission from the wind source to the population centers.

References

Bentek Energy, LLC (2010) *How Less Became More: Wind Power and Unintended Consequences in the Colorado Energy Market*, http://docs.wind-watch.org/BENTEK-How-Less-Became-More.pdf (accessed January 28, 2016).

Energy Information Administration (2013) *Capital Costs for Electricity Plants*, http://www.eia.gov/forecasts/capitalcost/(accessed January 26, 2016).

Energy Information Administration (2014) *Levelized Cost and Levelized Avoided Cost of New Generation Resources in the Annual Energy Outlook 2014*, Energy Information Administration, Washington DC.

Energy Information Administration (n.d.) *Electricity Data Browser;* http://www.eia.gov/electricity/data/browser/#/topic/0?agg=2,0,1&fuel=vtvo&geo=g&sec=g&linechart=ELEC.GEN.ALL-US-99.A~ELEC.GEN.COW-US-99.A~ELEC.GEN.NG-US-99.A~ELEC.GEN.NUC-US-99.A~ELEC.GEN.HYC-US-99.A~ELEC.GEN.WND-US-99.A&columnchart=ELEC.GEN.ALL-US-99.A~ELEC.GEN.COW-US-99.A~ELEC.GEN.NG-US-99.

A~ELEC.GEN.NUC-US-99.A~ELEC.GEN.HYC-US-99.A~ELEC.GEN.WND-US-99.A&map=ELEC.GEN. ALL-US-99.A&freq=A&ctype=linechart&ltype=pin&rtype=s&maptype=0&rse=0&pin= (accessed January 28, 2016).

Morvay, Z. K. and Gvozdenac, D. D. (2008) Applied Industrial Energy and Environmental Management, John Wiley & Sons, Chichester.

National Energy Technology Laboratory (2012) *Quality Guidelines for Energy System Studies – Detailed Coal Specifications*, http://www.netl.doe.gov/File%20Library/research/energy%20analysis/publications/QGESS_Detail CoalSpecs_Rev4_20130510.pdf (accessed January 27, 2016).

Natural Gas Supply Organization (2015) *Comparison of Fuels Used for Electric Generation in the US*, http://www. ngsa.org/analyses-studies/beck-data-rev/ (accessed January 27, 2016).

Speight, J. G. (ed.) (1997) *Petroleum Chemistry and Refining*, Taylor & Francis, Washington DC.

Case 9

Heat Exchangers and Drain Line Sizing

A client with a long relationship with your engineering company approached the heat transfer division requesting a solution to an operating problem on a heat exchanger. After discussing the problem with the client and analyzing it internally, the heat transfer division raised the problem with the chief engineer, who realized that the issue was related to the overall design of the exchanger system. The chief engineer called you to his office and explained that the system involves a nuclear steam turbine, a surface condenser, downstream exchangers, and interactions with the environment.

The system is complicated but the chief engineer is confident that you will be able to answer the client's questions and be prepared meet him within 1 week for an internal discussion of the answers to their questions and concerns. Following the internal meeting, you will have 1 week to modify your presentation for a meeting with the client.

The exchanger in question is the lowest pressure feedwater heater in a six-feedwater-heater cycle of a boiling water reactor (BWR) nuclear steam cycle. It is a shell-and-tube exchanger with feedwater flowing through the tubes heated by steam extracted from the steam turbine. Downstream of the turbine extraction, steam expands through one final stage of the steam turbine before exhausting into a steam surface condenser. In addition to extraction steam, the feedwater is heated by drains cascaded from higher pressure feedwater heaters. Condensed steam and cascaded drains exit the shell of the exchanger and are directed to the condenser with the steam-turbine exhaust.

Steam condensate with the cascaded drains exits the exchanger without a liquid level present in the shell. After exiting, the drain line carries the exchanger drains through a "gooseneck" pipeline and through a restriction orifice that meters the drains into the condenser.

According to the client, a liquid level appears in the exchanger during the winter months causing an alarm that forces the operators to decrease the thermal output of the nuclear reactor.

Case Studies in Mechanical Engineering: Decision Making, Thermodynamics, Fluid Mechanics and Heat Transfer, First Edition. Stuart Sabol.
© 2016 John Wiley & Sons, Ltd. Published 2016 by John Wiley & Sons, Ltd.
Companion website: www.wiley.com/go/sabol/mechanical

Without a decrease in the nuclear reactor thermal output, liquid might enter the steam turbine causing an immediate plant shutdown, potential damage to the turbine, and a possible release of radioactive material. The client has asked: "What is wrong?" and "What happened to cause the winter-time problems?" After these questions are answered, the client would like to explore a preliminary set of possible solutions to the problem.

The answer to the first question is a straightforward solution of energy and heat transfer equations with fluid dynamics. The second question is more subtle. Its answer requires working through the initial design calculations, which will not be that simple. The original energy balances were completed in the late 1960s with a mainframe computer. That computer has since been dismantled, and the Fortran software program no longer functions on personal computers used by your company. Even if you could locate the original Fortran code, converting it to a usable format would take much longer than the time available.

Recreating the original computer simulation model of the facility with a modern PC-based package could tie up too much of the available week. Therefore, you have decided to isolate the exchanger and condenser and approximate the energy-balance and heat-transfer calculations. The client has not requested a final solution – only an explanation of the problem, and you suspect that an approximation will be able to provide the answer. You will address the second question simultaneously with a request for the microfilm records through your contacts within the company.

You have worked with the client's engineers and know them to be capable, thorough, and experienced. You suspect from your conversation with the chief engineer that the client knows the answers to the first two questions, and is looking to your company for either confirmation or something else that you have not quite put into words. The client may also be seeking assistance from a competitive engineering firm.

9.1   Background

9.1.1   Steam Surface Condensers

Condensers for large steam-turbine generators are designed to discard waste heat from the steam power cycle to the atmosphere, most often to water or air. The engineer optimizes the capital cost and performance of the steam-turbine power plant to provide a system of pumps or fans, water piping, and the condensing equipment in a way that minimizes capital cost for the maximum steam turbine output. A wet surface condenser is a shell-and-tube design with cooling water on the tube side and steam condensing on the shell side. Cooling water may be supplied from a body of water, or from a cooling tower. Dry condensers are finned tube exchangers with steam on the inside of the tubes. Air is blown across the tubes to remove the latent heat from the steam.

The condenser's design balances two requirements: (i) the removal of latent heat from the steam given by equation (9.1), and (ii) the heat-transfer relationship shown in equation (9.2), where LMTD is given by (9.3) and its terms are defined in Figure 9.1. The flow from the steam turbine and its efficiency determine the mass flow ($\dot{m}$) and inlet enthalpy (h) to the condenser giving the required energy that must be transferred to the cooling water. The heat exchange area (A) and the exchanger's characteristic heat transfer coefficient (U) determine the saturation temperature (T_{sat}) of condensing; and thus the latent heat in the steam. The condensing

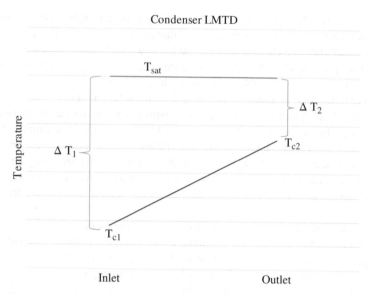

Figure 9.1 Condenser LMTD definitions.

pressure, and condensate enthalpy determined by equation (9.2) balance with the requirement to condense the quantity of steam at its enthalpy from the steam turbine – equation (9.1):

$$Q = \sum_{i=1}^{n} \dot{m}\Delta h_i \qquad (9.1)$$

where i = inlet stream flow 1 1 through n

$$Q = UA(LMTD) \qquad (9.2)$$

$$LMTD = \frac{-(T_{c2} - T_{c1})}{\ln\dfrac{(T_{sat} - T_{c2})}{(T_{sat} - T_{c1})}} \qquad (9.3)$$

For a wet condenser, the capital cost to install the cooling water piping, cooling water pumps condenser tubes and shell is optimized with the steam turbine output and cycle efficiency. As an example, a high rate of cooling water flow can decrease the condensing pressure, thus improving the steam turbine output but the piping capital cost, and pumping power soon overwhelm the improved steam-turbine output.

The inlet temperature of the cooling water from a water body is a boundary condition for the designer. This temperature changes with seasons, affecting the performance of the steam turbine and the design of related systems. If a cooling tower is used, its size is optimized with the cooling water, condenser, and power generation system to provide the owner with the best combination of capital and operating costs, with steam-turbine output.

Materials for a wet-surface condenser are selected for performance, reliability, and corrosion resistance. Copper alloys have good heat-transfer characteristics, and are generally

Figure 9.2   Steam surface condenser. *Source*: Reproduced by permission of Alstom.

resistant to most corrosion mechanisms in cooling water. However, these alloys do not have the strength to tolerate high tube velocities and thin walls. Stainless steels, or titanium alloys, are more costly than copper alloys, and suffer from lower heat-transfer characteristics but these materials have the strength to allow thinned wall tubes that can have high water velocities. Higher tube velocity improves heat transfer, and physically scours the tube to remove scale that builds up at speeds suitable for copper-alloy tubing. Therefore, a condenser with titanium tubes is smaller, lighter, and can cost less than a copper alloy design, even with the high-cost material.

The tube bundle of the wet condenser is designed to guide steam around the full perimeter of the shell, and the steam flows radially towards the center. Steam flowing from the bottom serves to reheat droplets falling from tubes in the upper section of the condenser. This reduces subcooling of the condensate below the condensing temperature and improves the efficiency of the exchanger. See the tube bundle design shown in the cutaway of Figure 9.2. Noncondensable gases that may be present in the steam collect in the center of the bundle and are evacuated.

In a steam-turbine power plant with multiple low-pressure steam turbines, the condenser is often a multipressure design. The cooling water flows continuously from one condenser to the next, resulting in a higher condensing pressure in the downstream condensing sections. Steam condensate from a lower pressure condenser section flows to the higher pressure section through a difference in static head (see Figure 9.3). Steam is directed under the distribution platform and is condensed by direct contact with the colder condensate, reheating the condensate to the higher pressure condenser saturation temperature. A multipressure design improves the capital cost and use of resources.

For an air-cooled, or dry, condenser the air flow rate and finned surface area are optimized with steam-turbine performance in much the same way as for a wet condenser. Dry condensers are considerably larger than a wet design due to the lower coefficient of heat transfer

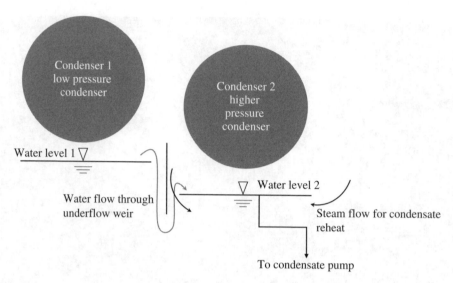

Figure 9.3 Steam condensate flowing to the higher pressure section through a difference in static head.

Figure 9.4 Power plant with an air-cooled condenser. *Source*: Reproduced by permission of the ATCO Group.

between the air and tube surface. Optimization of an air-cooled condenser results in a higher condensing pressure than for a wet condenser. Due to their high cost, air-cooled condensers are used only when conditions prevent using water-cooled designs. For example, when there is a shortage of water, or where environmental considerations limit temperature rise or discharge of dissolved solids from a wet cooling tower, an engineer may specify an air-cooled design. In Figure 9.4, the air cooled condenser is to the right of the main power station. Steam is directed into the condensing surfaces through the four parallel white ducts on the top of the condenser. Fans are located beneath the condenser, forcing air through the unit, enhancing heat transfer.

9.1.2   Feedwater Heaters

Feedwater heaters function like condensers – equations (9.1) and (9.2) must be satisfied for the energy and heat transfer within the exchanger. Most are a shell-and-tube design with feedwater on the tube side and steam condensing on the shell side. An example feedwater heater is shown in Figure 9.5. In this example, there are three steam-extraction nozzles entering along the top of the exchanger. Feedwater would enter through the lower nozzle in the hemispherical head shown at the left, and leave through the upper nozzle. There is an inspection port shown in the feedwater inlet.

In the solution of equations (9.1) and (9.2), feedwater heaters differ from condensers in that shell pressure is determined by the flow through the turbine stage following the steam extraction. The energy equation (9.1) is balanced with the heat transfer equation (9.2) by matching quantity of steam extracted from the turbine with the temperature rise of feedwater though the exchanger. The extraction pressure is updated with each iteration of the solution to reflect the steam flow to the following stage of the steam turbine.

The lowest shell-side pressure feedwater heaters are often located in the "neck" of the condenser – the transition zone between the steam turbine exhaust and the condenser tube bundle. Figure 9.2 shows three parallel condensers with feedwater heaters located in the neck of each. These exchangers operate at a pressure less than atmospheric and their location directly beneath the turbine simplifies piping designs.

Often in large power plants, there are many stages of feedwater heating. Feedwater heater drains are cascaded from higher to lower pressure exchangers. Drains from the lowest pressure exchanger may be pumped ahead into the feedwater, or simply routed to the condenser where any usable energy would be discarded to the atmosphere.

Figure 9.5   Alstom feedwater heater. *Source*: Reproduced by permission of Alstom.

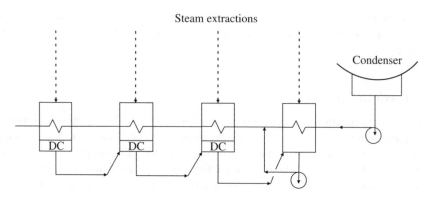

Figure 9.6 Low-pressure feedwater heater with cascaded and pumped ahead feedwater heater drains

Figure 9.6 shows a typical arrangement of four low-pressure feedwater heaters. The lowest pressure heater has drains that are pumped forward into the feedwater stream. The following exchangers, in the direction of feedwater flow, each has a separate drain cooler (DC) section in addition to the condensing section within the heater, and the drains cascade to the next lowest pressure exchanger.

9.1.3   Overall Heat Transfer Coefficient

The overall heat-transfer coefficient U of equation (9.2) for a shell-and-tube exchanger can be calculated based on the tube inside (U_i) or outside U_o area with the corresponding area, inside or outside, used to complete the equation. The heat transfer coefficient contains terms for internal and external film coefficients and the thermal conductivity of the tube material as shown by equation (9.4):

$$U_o = \frac{1}{\dfrac{A_o}{A_i}\dfrac{1}{h_i} + \dfrac{A_o}{L}R_s + \dfrac{1}{h_o}} \tag{9.4}$$

The internal, external and conduction components of U for condensers and feedwater heaters are described below in sections 9.1.4 through 9.1.6. Understanding how these coefficients contribute to the overall heat-transfer coefficient, is helpful when selecting heat transfer equipment, and in designing systems that require heat transfer. It is essential for this case study.

In many engineering applications, U is determined by the heat exchanger manufacturer, who often employs proprietary functions and methods for the internal and external film heat transfer coefficients. Those formulations may also include owner-supplied, or experienced-based fouling factors in addition to the manufacturer's margin to ensure performance guarantees are met. It can therefore be difficult to duplicate a manufacturer's overall heat-transfer coefficient from the basic equations, shown for condensers and feedwater heaters in sections 9.1.4 through 9.1.6 below. The value of U is sometimes provided in an equipment data sheet. If not, it can be determined from the guaranteed performance data on the exchanger data sheet using equation (9.2).

9.1.4   Condensing Heat Transfer

The condensing film coefficient on the outside of a horizontal tube generally provides little resistance to the flow of heat. The heat transfer coefficient for condensing on the outside of tubes from Nusselt (Holman, 1976: 358) is shown in equation (9.5):

$$h_o = 0.725 \left[\frac{\rho\left(\rho - \rho_g\right)gh_{fg}k_f}{\mu_f d\left(T_g - T_w\right)} \right]^{1/4}$$ (9.5)

Values of h_o for low-pressure steam surface condensers are in the neighborhood of 30 000 W/m²/K (5,200 Btu/h/ft²/R).

9.1.5   Forced Convection Inside Tubes

Heat transfer coefficients for water flowing inside tubes are typically 20 to 30 times less than for condensing heat transfer, and most often the controlling parameter for the overall heat transfer coefficient in water cooled condensers and feedwater heaters. For turbulent flow inside tubes, the coefficient can be found from equation (9.6) (Holman, 1976: 204):

$$h_i = 0.023\frac{k}{d}R_d^{0.8}Pr^{0.4}$$ (9.6)

This can be rewritten as shown in equation (9.7):

$$h_i = 0.023\frac{k}{d}\left(\frac{\rho Vd}{\mu}\right)^{0.8}\left(\frac{c_p\mu}{k}\right)^{0.4}$$ (9.7)

9.1.6   Conduction Heat Transfer

Resistance to heat transfer tubes per unit length is calculated using equation (9.8) (Holman, 1976: 388). For thin-walled stainless-steel or titanium tubes, the resistance due to conduction through the tube is about 45% of the thermal resistance due to the internal film coefficient.

$$R_s = \frac{ln\left(r_0/r_i\right)}{2\pi k}$$ (9.8)

9.1.7 Off-Design Exchanger Performance

Often an engineer is asked to provide, or necessity dictates, an estimate of an exchanger's performance at operating conditions that are different from those on an equipment data sheet. Boundary conditions of temperature or pressure may change due to ambient conditions, process output level specified by different design cases, or modifications to the process in which the exchanger is located. There may also be changes to the fluid properties due to a change in a feedstock, implementation of a new chemical process, a change of catalyst, and so forth. Determining the new performance characteristics of the heat exchanger can be overwhelming without a few tools and short cuts.

Commercially available software packages can make this type of calculation. The package may be devoted to heat exchangers, or the heat transfer equations may be imbedded within a modeling program that includes thermodynamic and transport properties for mixtures and pure substances. When using a modeling software package, it is often reasonable to determine a value for the heat-transfer area (A) using the package's formulation of U at the manufacturer's guaranteed conditions. Once determined, the area is a constant in equation (9.2), and the software package can be used to predict the heat exchanger's performance at new process conditions. Reasonable assumptions for the numerous parameters necessary to determine the heat transfer area (fin height, length and density, diameters, materials, lengths, etc.) can provide an accurate basis for predicting off-design performance.

In the absence of a computer software package, evaluating U at off-design conditions can be burdensome, especially in cases where high accuracy is not required. However, an understanding of the basic formulas above can provide acceptable approximations. In the example of a shell-and-tube condenser, the overall heat transfer coefficient, U, is determined mostly by the internal film coefficient. Due to its comparative value, changes in the resistance of the condensing film can be ignored and the thermal resistance of the tube can be approximated by a constant. Therefore, for no change in the fluids exchanging heat, and moderate changes in process conditions, U can be approximated as being proportional to the internal film coefficient, as shown in equation (9.9):

$$U \propto h_i \tag{9.9}$$

For small process changes in a steam condenser with water on the tube side, the thermal conductivity, specific heat, and density of the cooling water are essentially constant. The internal heat transfer coefficient, equation (9.7), therefore becomes a function of the mass flow rate and viscosity. Thus, the overall heat-transfer coefficient can be approximated with equation (9.10):

$$U \propto \left(\frac{\dot{m}^{0.8}}{\mu^{0.4}} \right) \tag{9.10}$$

Results that require a high level of accuracy should rely on qualified software and rigorous calculations of heat transfer coefficients. When an approximation is sufficient, determine the design value for UA from equation (9.2) at the manufacturer's guarantee

Figure 9.7 Pouring with Froude number greater than 0.3.

conditions. Then, approximate *UA* at various conditions using an equation developed in a manner similar to equation (9.10) that is derived for the specific type of heat exchanger being studied.

9.1.8   Drain Line Sizing

There are two very broad categories of drain lines: those where there is a liquid level in the vessel, and those where there is not. Occasionally, it is impractical to hold a liquid level in the vessel being drained and for those designs, the drain line is sized conservatively large to allow vapors to flow countercurrent to the liquid leaving the vessel. Perry and Green (1984: 5–44) report that at a Froude number less than 0.31 vapor is not entrained in a liquid flow, allowing the two phases to flow in opposite directions. In practical designs for "self-venting" drain lines the pipe diameter is sized such that the Froude number is less than 0.3.

If the Froude number is greater than 0.3 (see Figure 9.7), flow from the vessel behaves much like soda, or beer being poured quickly from a long-neck bottle. The liquid flow rate is high enough to prevent a smooth flow of vapor and liquid in two directions. The Froude number is defined by equation (9.11):

$$Fr = \frac{V}{\left(g \cdot d\right)^{0.5}} \tag{9.11}$$

9.2   Reading

Read Sloley (2013) and answer the following questions.

Questions

1. Cooling water should be on the tube side of a steam surface condenser for the following reason(s):
   (a) Lower cost
   (b) Easier to clean
   (c) Freeze protection
   (d) Reduced steam side pressure drop
   (e) A, B and D above
2. Usually the high-pressure fluid is on the exchanger shell side. True or false?

9.3   Case Study Details

9.3.1   Flow Diagram and Equipment

The nominal 800 MW nuclear power plant in question has six stages of feedwater heating, numbered from the highest extraction pressure to the lowest. The later stages of the steam turbine have moisture removal with the extraction steam to improve the efficiency of the following expansion stages. The extraction steam quality is therefore not indicative of flow to the following steam-turbine stage group.

The flow diagram of the last two steam-turbine stage groups, condenser and feedwater heater, is shown in Figure 9.8. As shown, condensate from the condenser is heated against steam extracted from the steam turbine. Drains from the next higher pressure feedwater heater

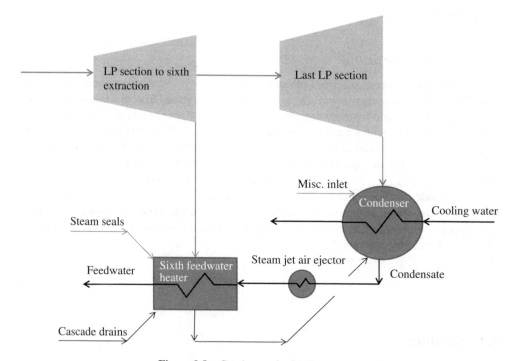

Figure 9.8   Condenser, feedwater system.

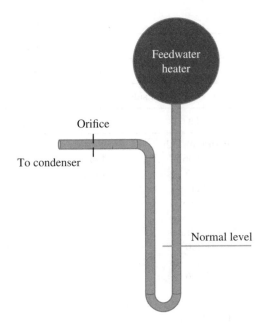

Figure 9.9 Drain line configuration.

enter the sixth heater in the cycle with extraction steam and turbine steam seals. Drains from the feedwater heater are cooled within the condenser and form part of the condensate to the feedwater heater.

As stated earlier, the feedwater heater in question operates without a liquid level. During normal operation, a level appears in the drain line some distance below the exchanger. An orifice in the 500 mm (20 in.) goose-neck drain line (see Figure 9.9) ensures that the feedwater operates at a higher pressure than the condenser. The liquid level in the drain line is monitored continuously in the control room with high and low liquid-level alarms in the event of abnormal operation. The backup high-high level indication and alarm instrument is located at the feedwater heater elevation.

The client reports that, during the winter months, high-high level alarms are triggered without first receiving a high-level alarm from the normal level transmitter.

The condenser is a single-pressure, shell-and-tube design with cooling water provided from an adjacent body of water. Cooling water passes once through the tube bundle and returns to the body of water. Inlet water temperature ranges from a minimum of 273.7 K to a maximum of 295.3 K with an annual average of 288.2 K. A portion of the condenser data sheet is shown in Table 9.1, and a portion of the feedwater datasheet is shown in Table 9.2 at 100% reactor power and annual average conditions.

9.3.2 Design Cases

In the design of essentially any piece of equipment, facility, or integrated set of projects, the engineer must consider how the item is to be used and under what set of circumstances. Rarely does a single set of design conditions dictate every design aspect of a system of components.

Table 9.1   Condenser data sheet.

Exchanger	Performance for One Unit			
	Surface condenser			
	Shell side		Tube side	
	Inlet	Outlet	Inlet	Outlet
Steam inlet	LP Turbine Exh.	Condensate		
Flow (kg/s)	297.56	490.8		
Pressure (kPa)	4.826	4.826		
Temperature (°C)	Sat	Sat		
Enthalpy (kJ/kG)	2209.5	Sat		
Liquid inlet	Cascade drains		Cooling water	
Flow (kg/s)	118.6		16,583	
Temperature (°C)	Sat		15	25
Pressure (kPa)	4.826		138	69
Enthalpy (kJ/kg)	309.8			
Other inlet	Misc. drains			
Flow (kg/s)	74.6			
Enthalpy (kJ/kg)	886.7			

Table 9.2   Feedwater heater datasheet.

Exchanger	Performance for one unit			
	Sixth extraction feedwater heater			
	Shell side		Tube side	
	Inlet	Outlet	Inlet	Outlet
Steam inlet	LP turbine ext.	Condensate		
Flow (kg/s)	46.1	118.6		
Pressure (kPa)	38.8	38.8		
Temperature (°C)	Sat	Sat		
Enthalpy (kJ/kg)	1999.2	Sat		
Liquid inlet	Cascade drains		Condensate	
Flow (kg/s)	72.1		490.8	490.8
Temperature (°C)	Sat		32.8	71.2
Pressure (kPa)	38.8		1896	1758
Enthalpy (kJ/kg)	319.4			
Other Inlet	Steam seal regulator			
Flow (kg/s)	0.33			
Enthalpy (kJ/kg)	2586.5			

One set of conditions may dictate the size of a piece of pipe, for example, while others dictate its thickness, support structure, material of construction, type of paint, insulation, and so forth. Equipment used for cooling, a condenser for example, may be designed around the worst case summer conditions, while the design of adjacent and downstream equipment may be determined by the extreme winter conditions.

In order to ensure a design is adequate for all conditions, the owner (purchaser) and the engineer begin the design process by defining a set of design cases. Once determined, the owner specifies the cases in the contract under which the engineer provides services. Importantly, the engineer must notify the owner, if a new case develops during the design process, to ensure the finished product is adequate, safe, and reliable under the anticipated operating modes, range of feedstock characteristics, and environmental conditions.

In the case of a nuclear power plant, the safe design must consider potential operation at a margin above the licensed reactor power limit, winter and summer conditions, as well as operation during startup and stable minimum load. For this case study, assume the design must be adequate at 102% of the licensed reactor power. Also assume that feedwater flow is proportional to the reactor power level.

9.3.3   Exercises

The following exercises take you through the basic steps of this case study – application of fundamentals, analysis of the problem, and development of an approach for the client. You are encouraged to complete the calculations with a spreadsheet rather than a custom software package.

Part 1: Application of Engineering Fundamentals

1. Using the data in Table 9.2, calculate the required drain-line size for the sixth feedwater heater so that there is no liquid level. Select the next largest standard pipe size. Use the "standard" schedule.
2. What error(s) occurred in completing 1 above?
3. Using data in Table 9.1 and Table 9.2 calculate the LMTD for the condenser and feedwater heater.
4. Using data in Table 9.1 and Table 9.2, calculate the product *UA* in equation (9.2).
5. Solve equations (9.1) and (9.2) for the condenser and feedwater heater during the winter conditions. Use the goal-seek function of a spreadsheet software, or manually iterate, to find the condensing pressure and the feedwater temperature leaving the feedwater heater such that Q in the two equations is equivalent for each heat exchanger. Make the following assumptions:
   (a) The steam turbine extraction pressure is proportional to the square of the flow to the following stage.
   (b) The feedwater temperature rise across the steam jet air ejector condenser is a constant.
   (c) Reactor core power = 102% of the licensed core power.
   (d) Assume the flows of feedwater, miscellaneous drains to the condenser, and cascaded drains from the downstream feedwater heater are proportional to the reactor core power. The enthalpies of the miscellaneous and cascaded drains are constant.

(e) Assume the steam seal flow and enthalpy are constant. The author suggests entering the feedwater heater drain enthalpy as a constant in equation (9.1) to keep the goal-seek function stable. Manually change the value as necessary to achieve a converged solution.

6. Recalculate the feedwater heater drain line size as in 1. above. Select the next largest pipe size using the standard pipe schedule.

7. Calculate the Froude number at the winter time conditions using the pipe diameter selected in 1. above and compare it to the design requirement.

8. Why does a liquid level appear in the feedwater heater in the winter?

Part 2: "What happened to cause the winter time problems?"

9. What do you suspect happened to cause the feedwater level problems of this case study?
10. What engineering processes could have prevented the line sizing error of this case study?
11. What owner process would have resolved the problem?
12. List two possible solutions for the drain line problem. List the advantages and disadvantages of the alternatives, and an engineering reason the proposed solution would function as expected.

Part 3: Client Interface

13. As a representative of the engineering firm, what would be your approach to answer the client's two questions: What is wrong and what happened? (Open discussion.)
14. As the owner, what would be most important aspect of the engineer's answers? (Open discussion.)

9.4   Closure

This case study explores three levels of engineering: the application of fundamentals, engineering processes, and client interfaces. The fundamentals in this example bring together heat transfer, and fluid dynamics in the design of a vessel drain. The engineering processes involve quality assurance and quality control measures for engineering firms, as well as a purchaser of engineering services. When presenting information to a client, an engineer should take into account what the client finds important. This aspect of engineering can be difficult, as each client may determine value differently. It is important, as in this case, to carefully read the question, and understand the client's motivations, and if possible, the hidden meanings. In all cases, one value stands apart from all others – integrity. Maintaining the highest integrity on behalf of your company and yourself, even when admitting an error, pays dividends.

Microfilm records of the calculations for feedwater heater drain line size showed that the calculation was not signed or dated by the engineer. It was not checked by a peer, or supervisor. The calculation was performed only once for the average ambient conditions, which, as shown by the exercises of this case study, was not the governing case for the size of the drain line. During the winter months, colder condensate from the condenser lead to higher steam extraction flows, which overwhelmed the self-venting drain section and resulted in a water level inside the heat exchanger. Continued operation and full power would have partially filled the exchanger, potentially leading to unstable operation and the possibility that water could be introduced into the steam turbine.

The owner accepted the facility and allowed the warranty period to expire without raising the issue to the engineer. When approached for the solution, the engineer had no liability for the initial error, and was in position to charge the client a second time for the work originally performed. After recognizing its error, the firm offered to complete a new design in cooperation with the client, without charging for engineering labor and expenses. The client accepted the offer, extending a strong relationship with the engineer, and the firm made a profit on the equipment, materials and installation of the new design.

The series of errors of this case were largely related to quality control and assurance. Between the time when the calculation was performed and the start of the case study, quality management in engineering companies across Europe and the North America had become a normal part of business, driven, in part, by competition with Japan. Prior to the 1970s, a corporation relied on individuals for quality assurance rather than a set of written practices and procedures. By adopting a corporate policy on quality management, and developing procedures with specific responsibilities through the company, engineering deliverables became more consistent with fewer errors. Finished products and projects are now safer and more reliable with fewer initial problems and fewer warranty claims. Engineering corporations can rely on improving sets of procedures rather than an individual to uphold the company's reputation.

Owners too, have become more involved in the initial design and work products. Defining the initial design requirements in cooperation with engineering firms, requiring specific design cases that must be analyzed prior to acceptance of a design, and keeping the project team engaged through the warranty period have steadily improved schedule, cost, and the quality of the completed projects.

In this study, the engineer proposed pumping the fifth extraction heater drains ahead of the heater into the feedwater system. That proposal replaced sensible heat transferred from condensate drains to latent heat from condensing steam, which resulted in slightly more heat transfer to the feedwater with a lower drain flow. The existing drain line was more than adequate for the new design, and winter level problems were eliminated. The new design also made a marginal improvement to the overall steam cycle efficiency.

In discussions with the client, recognition of the original engineering error translated into a selling point for the engineering firm, whereby the firm was able to describe their quality assurance programs and improvements since the late 1960s. It turned out that the owner was fully aware of the cause for the level problems and was seeking an answer from the engineering firm related to their quality-assurance practices. By presenting the findings in a straightforward, professional manner together with a description of the QA procedures, the firm earned a chance to correct the problem.

9.5   Symbols and Abbreviations

c_p:	constant pressure specific heat (kJ/kg·K);
d:	diameter (m);
Fr:	Froude number (eq. 9.11);
h:	enthalpy (kJ/kg);
h:	heat-transfer coefficient (W/m²·K);
LMTD:	log mean temperature difference (K) (eq. 9.3);
k:	thermal conductivity (W/m·K);
Nu:	Nusselt number (h·x/k);
P:	pressure (kPa);
Pr:	Prandtl number (c_p·μ/k);

Q: heat transfer rate (kJ/unit time);
R_d: Reynolds number based on diameter ($\rho V d/\mu$);
T: temperature (°C, K);
V: velocity (m/s);
x: length dimension (m);
Δ: change operator;
μ: dynamic viscosity;
ρ: density (kg/m³).

Subscripts

1, 2, ... :   sequential position;
c: cold fluid;
h: hot fluid;
sat: saturated conditions;
f: saturated fluid.

9.6   Answer Key

Section 9.2: Reading

1. e.
2. False.

Section 9.3.3

Part 1:

1. Line size: 500 mm, d = 489 mm.
2. All design cases have not been considered.
3. Condenser LMTD = 11.5 K:
   (a) Feedwater heater LMTD = 16.1 K.
4. Condenser UA = 60 175 kW/K.
   (a) Feedwater heater UA = 4897 kW/K.
5. See Figure 9.10.
6. Line size: 600 mm, d = 590.6 mm.
7. Fr = 0.33 10.2% error.
8. With winter cooling water temperatures, the self-venting feedwater heater drain line is overwhelmed by the liquid flow from the exchanger.

Part 2:

9. The most likely cause for the incorrect feedwater heater drain line size was a failure to check the design for all operating cases. The size calculated by the case study exercises should be checked again with a conservative manufacturer margin for the condenser and feedwater heater surface area.
10. An engineering quality assurance program that included the following may have prevented the incorrect sizing:
    (a) A corporate quality assurance policy.
    (b) Engineering quality assurance procedures with assigned responsibilities within the corporate organization structure.

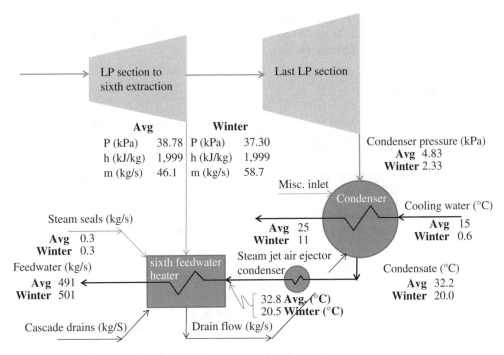

Figure 9.10 Answer: section 9.3.3, number 5.

(c) A written document outlining the design cases reviewed by the client.

(d) A design procedure that required multiple calculations, one for each design case.

(e) A document listing each calculation.

(f) A requirement to have calculations checked by a qualified peer or supervisor, with a recorded signature and date of the reviewer.

(g) A requirement to sign and date each calculation.

(h) An internal check of the design drawings against the approved calculations with signatures and dates of the reviews.

11. An owner project team, including a project manager, project engineer, operations, and maintenance representatives that remained with the project through the warranty period. The project team should ensure that any punch-list items identified at the completion of the project, and warranty issues that develop, are resolved to the satisfaction of the owner.

12. A reduction of liquid flow out of the feedwater heater can be accomplished with either a bypass of cascaded drains to the condenser or pumping the cascaded drains ahead to the downstream feedwater. These methods replace sensible heat transferred by the drains with latent heat from the extraction steam, which reduces the amount of drain flow from the feedwater heater. The bypass option decreases the cycle efficiency by rejecting more heat in the condenser. The option to pump the drains ahead increases the cycle efficiency. Alternatively, the heater drain could be replaced with the proper size. This option would retain the original design intent but would require a redesign of the condenser internals, work in a confined space, and a long replacement outage. The condenser may be

contaminated with radioactive material from the reactor core, creating an unsafe condition for the repair crews. Therefore, this case should be considered only if there are no other practical alternatives.

Further Reading

Cross, V. (n.d.) *What Are the Standard Elements of a Quality Management Plan?* http://smallbusiness.chron.com/standard-elements-quality-management-plan-22015.html (accessed January 30, 2016).

Dickey, J. B. Jr. (1978) *Managing Waste Heat with the Water Cooling Tower,* 3rd edn., The Marley Cooling Tower Company, Mission, KS.

Haslego, C. and Polley, G. (2002) Designing plate and frame heat exchangers. *CEP Magazine* (September), 32–37, http://people.clarkson.edu/~wwilcox/Design/hxdesign.pdf (accessed January 30, 2016).

Hawkins, G. B. (2013) *Overflows and Gravity Drainage Systems,* http://www.slideshare.net/GerardBHawkins/overflows-and-gravity-drainage-systems (accessed January 30, 2016).

MyCheme (2013) *Designing for Gravity Flow,* http://www.mycheme.com/designing-for-gravity-flow/ (accessed January 30, 2016).

References

Holman, J. P. (1976) *Heat Transfer,* 4th edn., McGraw-Hill, New York, NY.

Perry, R. H. and Green, D. (1984) *Perry's Chemical Engineer's Handbook,* 6th edn. (50th anniversary edition). McGraw Hill, New York, NY.

Sloley, A. (2013) Shell-and-tube exchanger: Pick the right side. *Chemical Processing,* http://www.chemicalprocessing.com/articles/2013/shell-and-tube-heat-exchanger-best-practices/(accessed January 30, 2016).

Case 10

Optimized Maintenance*

This case study returns to the steam-turbine power plant of Case 1, which is related to one power plant within a small fleet of generating assets recently purchased by an investment group. The aging, nominal 600 MW steam-turbine generator has lost 3.5% of its output and the cycle heat rate has increased about 0.7% in a little over 7 months. In addition to the loss of performance, the cost of natural gas has decreased in recent years, increasing market pressure on coal-fired generation, and an economic downturn has led to lower electric power demand in many of the company's service areas. Industries have idled factories, and the outlook is not promising for their return to production in the near future. These factors have created a challenging situation for the young investment company, which is now seeking ways to cut costs and improve returns for shareholders.

The management committee, set up by the investor group, governs the assets, hires the senior staff, provides guidance, and controls their budgets. Individuals from the investor group with financial, management, legal, and accounting expertise are members of the committee. At the direction of the board of directors of the investment group, the management committee has set up a task force charged with proposing alternatives that can reduce costs, and improve the health of the power-plant investments. Alternatives to be investigated include fuel switching, staff reconciliation, optimizing maintenance, asset sales, and closures.

The words "staff reconciliation" and "optimizing" used in the context of plant closures often connote a reduction in personnel. These are often the right and proper methods to improve the health of a corporation. Selling assets can unload a difficult or loosing portion of a portfolio but must often be done at a loss. Difficult times require difficult measures but

---

*This case study is based on Sabol (1985).

Case Studies in Mechanical Engineering: Decision Making, Thermodynamics, Fluid Mechanics and Heat Transfer, First Edition. Stuart Sabol.
© 2016 John Wiley & Sons, Ltd. Published 2016 by John Wiley & Sons, Ltd.
Companion website: www.wiley.com/go/sabol/mechanical

creativity is an asset within the company that can turn a situation around, prevent a sale, improve returns, and lessen the impact of a reduction in force.

Your supervisor was asked by the management committee to lead the maintenance task force and he has assembled a small team of engineers with a dual purpose of (i) determining the root cause for the loss of performance, and (ii) optimizing maintenance. You have been assigned the task of optimizing the maintenance systems in an effort to reduce costs.

Your first action is to visit the site with the "root-cause" team, interview the operations and maintenance staff and collect as much data as practical. You are able to collect results from a series of regular turbine performance tests, and have had time to discuss operating and maintenance practices at the site. The history is mostly since the time the investor group purchased the facility but some of the operators have worked there since it was constructed, and were able to discuss the full facility history.

You have a unique opportunity to effect the direction of the team's recommendation to the management committee and potentially prevent the sale or closure of the asset.

10.1 Background

10.1.1 Maintenance Practices

Maintenance protocols may be based on reactionary, preventive, condition or reliability-centered principals. *Reactionary maintenance* is the repair of equipment following a failure – fixing a flat tire for example. For critical equipment, especially for a large steam-turbine generator, repairing only after a failure has occurred, while at times necessary, is the least efficient and most costly maintenance protocol. Reactionary maintenance practices can result in uncertain production, loss of warranty provisions, and extensive collateral damage following a rotating or stationary component failure. Most large steam turbines are maintained on a *preventive* basis.

Preventive or time-based maintenance is a maintenance program that performs repair on a set schedule – often proposed by the manufacturer. The schedule may be based on the manufacturer's research and field experience that yields a schedule based on the fleet average and not the specific machine. The manufacturer's schedule would also be developed with its interests as a priority, and those may be quite different from the owner's. As an example, the manufacturer may propose a conservative repair cycle so that equipment failures are rare and their equipment performs better than their competition in terms of reliability. The owner might be willing to accept a less conservative repair period in order to increase production and reduce maintenance, especially in the case where experience shows a longer repair cycle is possible.

Condition-based maintenance (CBM) is a flexible maintenance regime based on measurements of the condition of a component, equipment item, or system. Measurements such as vibration, temperature, lubricating oil condition, or process parameters are monitored and interpreted so that maintenance is only performed when required. Condition-based maintenance has the potential to reach the minimum amount of maintenance without failure but there are drawbacks. Condition-based maintenance instrumentation systems can be expensive. The results require accurate interpretation and judgment. Interpretation cannot be perfect, so setting of the maintenance period requires a measure of judgment as much as the science behind the method.

Reliability-centered maintenance (RCM) is a maintenance practice developed by executives of the airline industry and currently defined by the technical standard SAE JA1011. It was originally established to institute the minimum required level of maintenance to preserve

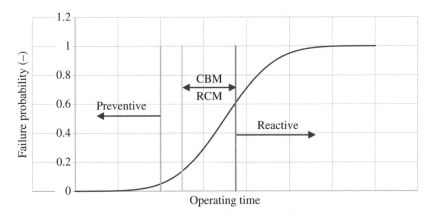

Figure 10.1   Maintenance practices.

system function. It was intended to be a complete maintenance protocol including failure analysis, and analysis of equipment requirements, results of a failure, programmatic processes that prevent failures, and so forth.

When changing from a preventive, time-based maintenance program recommended by the equipment manufacturer or the maintenance services provider to the CBM or RCM, it is important to consider any additional risks the owner would have to accept. Those risks may include the loss of warranty coverage if the time between overhauls exceeds contract limits. Risks can also include loss of collateral damage coverage, reliability guarantees, and so forth. Therefore, careful consideration of contact details, where appropriate, must be an integral part of an evaluation of maintenance practices.

Condition-based maintenance can be impractical when there is a long planning period required for an overhaul. For example, the time it takes to prepare for a refinery or chemical plant turnaround can exceed several months. Waiting until a critical equipment item requires immediate repair would cause unnecessarily high labor and parts costs, and could result in uncertain product delivery. In addition, seasonal pricing patterns often dictate that major maintenance occurs during a certain time of the year. Therefore, a corporation often prefers time-based maintenance for large complex systems, those with seasonal pricing pressure, or where there are other risks imposed by a condition-based system.

Figure 10.1 illustrates the four types of maintenance protocols mentioned above on a probability chart. Reactive maintenance would occur the least often of the three methods and would maximize the operating period between repairs. Preventive maintenance would be the most conservative, occurring on the shortest cycle, which would be predetermined based on experience or a set of criteria. Condition-based maintenance and RCM would likely occur somewhere between preventive and reactive maintenance. With CBM, some cycles would be shorter than others depending on how the equipment or system behaved during the cycle.

10.1.2   Economic Model for Maintenance

The three maintenance practices discussed in section 10.1.1 have in common the assumption that they are optimized for the type of equipment or company performing the maintenance.

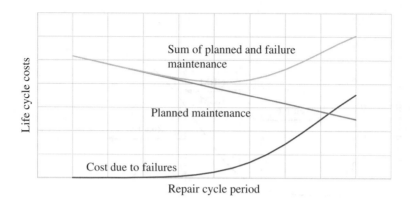

Figure 10.2   Life cycle maintenance costs.

All three may be part of a maintenance system and applied depending on the equipment size, cost of repair, or impact of a failure on the economic performance of the company. Figure 10.2 is an example of an optimization that accounts for the probability of a failure, the cost of the failure, including the lost production during maintenance, and the cost of normal repair prior to the failure. If properly optimized, the company would operate with a maintenance interval that minimizes the total lifecycle cost of maintenance.

The model utilizes the normal probability distribution, with the cost of repair, and an assumed standard deviation. Programs such as CBM have an underlying objective to increase the certainty when a part may fail so that maintenance intervals can be extended without undue risk effectively decreasing the standard deviation. Due to the interpretive nature of CBM programs, however, uncertainty of the moment of failure cannot be completely eliminated; and a measure of conservatism must be built into the decision process to avoid unexpected events.

Comparatively, the mean time to failure and standard deviation are known with greater certainty with CBM than with preventive maintenance but both programs have similar objectives, and can follow a similar economic model.

10.1.3   Operating Costs other than Maintenance

Much of the literature and research devoted to preventive maintenance and CBM focuses on extending run time and repairing equipment prior to a failure. However, preventing the added cost that results from a failure, which may include increased outage durations, additional parts replacement, expediting fees, unstable operations, and so forth, does not capture increases in variable operating costs due to performance degradation.

For cases in which lost performance is significant when compared to the cost of maintenance, economic optimization should account for changes in output and the cost of production. Systems that can be impacted by performance degradation include those with compressors and heat exchangers that can become fouled, long-distance piping systems prone to scale or sludge buildup, turbines exposed to impurities, and so forth. Whenever a process requires periodic cleaning, or the repair of worn parts, performance impacts on operating costs could be an important consideration in addition to the avoidance of a failure in a maintenance program.

10.2   Refresher

10.2.1   Cost to Generate Power

The cost to generate electric power is calculated from: (i) the price of fuel per unit of energy value of the fuel, $/GJ, or $/MMBtu, or pence/therm, for example; (ii) the heat rate of the power generating station plus (iii) nonfuel fixed, and (iv) variable, operating costs. The heat rate is expressed in terms of the energy input from fuel required per MWh of production: kJ/ kWh or Btu/kWh. The fuel cost multiplied by the heat rate, and converting units, yields the cost of fuel to produce electricity, usually in price per kWh or per MWh. The units may be selected for convenience of the calculation or the intended audience.

10.2.2   Fixed and Variable Operations and Maintenance (O&M)

Fixed O&M are those costs that are required to retain the asset and do not vary with production. Fixed O&M costs for electric-generating plants are usually expressed in literature based on the capacity of the facility – $/kW, for example. In a comparative analysis between options for a single facility, the fixed component of O&M is often neglected as it does not change between options.

Variable O&M is generally expressed as a price per unit of output, MWh for a power plant. For a modern coal-fired power station, the nonfuel variable O&M is about $4.47/MWh (Energy Information Administration, 2013: Table 1). Comparative studies of options that impact output and efficiency must consider the variable component of O&M.

10.2.3   Cost of Fuel

Solid fuel is usually priced per unit of weight, most often short tons (2000 lb) in the United States and tonnes (1000 kg) elsewhere. For this study, assume a subbituminous coal with a higher heating value of 19 920 kJ/kg has a delivered price of $35.27/te. The cost per tonne divided by the heating value and multiplied by 1000 yields the price of coal in $/GJ.

10.2.4   Short-Run Gross Margin

When calculating a difference in earnings between operating cases, fixed operating costs, interest, depreciation, and amortization, charges are often assumed to be constant between the cases. Therefore, knowledge of only the variable revenues less variable costs, the short-run gross marginal cost, is sufficient to determine the difference in earnings between the cases.

10.3   Presentation Techniques

10.3.1   Waterfall Chart

A waterfall diagram is an effective visual tool to show the elements of a sum. Relative sizes for each component of the total, positive or negative, are easily discerned from the presentation. The visual interpretation improves the understanding of the material with minimal explanation.

It is a stacked bar chart constructed in Excel with two data sets. One set is shown on the chart with no color and no outline and is therefore "hidden." The other set is plotted with a pattern or color, and is "visible." The two data sets are sorted to show the elements of a progression from an initial state to a final state. Figure 10.3 is an example water flow showing three reasons for improved earnings between Quarter X and Quarter Y. Increases in sales volume and sales price provided increases to earnings of $125 000 and $36 000 respectively, while an unplanned outage resulted in a loss of $25 000. Positive and negative numbers are shown with different patterns or colors. The net change from Quarter X to Y, $136 000 is shown with a different pattern or color to represent the net result. Data for the chart is shown in Table 10.1.

To construct a waterfall chart, begin by constructing a table with the series and progressive total as shown in Table 10.1. The first row of figures (row 4 of the spreadsheet) has a zero value for the "hidden" series (column B), the actual value for "visible" series (column C), and the change in earnings is shown in column D. The next row of the table has a value for the hidden series equal to the sum of the previous row (row 4) columns B and C. Column C shows the increase in earnings due to price, $36 000. Column D shows the sum of columns B and C. Negative values must be tracked individually. The negative value due to the unplanned outage is

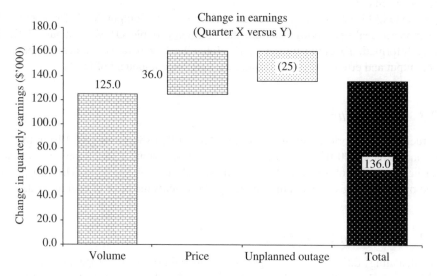

Figure 10.3   Waterfall chart.

Table 10.1   Example of waterfall chart values.

Column	A	B	C	D
Row		Hidden	Visible	Δ earnings
4	Volume	0.0	125.0	125.0
5	Price	125.0	36.0	161.0
6	Unplanned Outage	136.0	25.0	136.0
7	Total	0.0	136.0	136.0

Table 10.2   Example waterfall chart calculations.

Column	A	B	C	D
Row		Hidden	Shown	ΔEarnings
4	Volume	=D4–C4	125	125
5	Price	=SUM(B4:C4)	36	=B5+C5
6	Unplanned Outage	=D5–C6	=ABS(D6–D5)	136
7	Total	0	=D7	=D6

presented with a "hidden" series value equal to the previous row total (column D, row 5) minus the loss, and a "visible" series value equal to the absolute value of the loss (see row 6 of Table 10.1).

The "hidden" series on the chart includes values in column B, rows 4 through 6. The "visible" series (column C) includes values from rows 4 through 7. Colors or patterns of the bars in Excel are assigned by double clicking each bar and setting parameters from the spreadsheet options. Negative and positive values must have different colors for clarity. As many rows as required may be added with similar logic to that present in Table 10.1. Table 10.2 shows the calculations described above. Values in the table may come from different sections of the spreadsheet.

10.3.2   Line and Scatter Plots

In Microsoft Excel, "line" charts show the numeric values along the ordinate (y-axis) corresponding to categories shown along the abscissa (x-axis) of the chart. Categories are equally spaced along the x-axis. A Microsoft "scatter" chart shows the relationship of x and y numeric values, usually for engineering and scientific data. A scatter chart can be used to determine the relationship between two parameters, which is different from the Microsoft "line" chart. Be sure to use the appropriate type of chart when plotting data in Microsoft Excel.

10.4   Reading

Read Horne and Dreher (1963) and answer the questions below.

10.4.1   Questions

Figure 10.4 shows the impact of speed on the contact area of the surface of an aircraft tire.

1. At the annual inspection required for your automobile, the vendor recommends that you replace your tires. The tread has not quite worn to the wear bar molded into the tire; but the mechanic has made a case that your tire performance has decreased enough to warrant replacement. Are you getting good advice, or is the vendor simply trying to hit sales targets for the month?
2. A tire wear bar molded into the tire tread is a form of CBM. Tires worn to the wear bar should be replaced as driving with worn tires may result in a traffic violation, and possible fines, in addition to being unsafe. Condition-based maintenance can improve safety, and lessen the cost of repair, compared to preventive or time-based approaches, such as replacing tires after 50,000 km. As tires do not wear identically for all road conditions and driving habits, CMB replacements may result in lower operating costs. How does the mechanic's advice in 1. above differ from CBM?

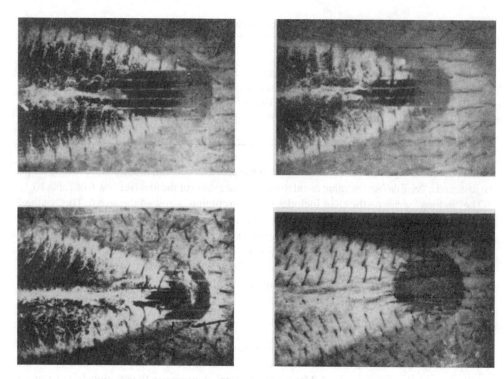

Figure 10.4   Aircraft tire on glass runway, 28 to 88 knots. *Source*: NASA.

10.5   Case Study Details

10.5.1   Data

The power-generating facility under study is a nominal 600 MW pulverized coal-fired power plant. The plant follows the manufacturer-recommended interval of 5 years for the steam turbine's major maintenance. Major maintenance includes complete disassembly of the turbine, removal and cleaning of the rotating and stationary parts, replacement of blades and vanes based on life and observed wear, and replacement of worn seal strips throughout the machine, including shaft seals. Minor repairs to stationary or rotating blades are often accomplished by blending, and polishing out defects with hand-held grinders.

The staff conducts regular turbine performance tests and these show the following trends for the first 7 months following a major overhaul:

- Measured high-pressure (HP) turbine efficiency fell about 1.2%.
- Measured intermediate pressure (IP) turbine efficiency shows little change.
- Calculated low-pressure (LP) turbine efficiency has fallen at a rate of 1.7% per year based on heat balance calculations.
- The turbine inlet flow passing ability has fallen at a rate of 14.4% per year.

To combat the losses in efficiency, the site conducts an HP turbine wash approximately every 9 months, about 280 days. Turbine washes are accomplished by temporarily operating the HP

Table 10.3 High-pressure turbine efficiency test results.

Days since overhaul	η_{HP} (%)
0	84.89
27	84.82
137	84.08
152	84.07
174	83.88
222	83.83
256	83.82
279	83.72
280	turbine wash
341	83.98
356	83.93

turbine with reduced pressure and temperature, forcing two-phase steam in the latter stages. After a wash, the HP turbine efficiency and inlet flow function partially recover. Analysis of the data shows the degradation trends of efficiency and flow function index to values seen 150 days prior to the wash, after which the trends continue in the same manner as prior to the wash.

High-pressure turbine efficiency data from the tests is summarized in Table 10.3. As shown, the water wash partially restored efficiency but the downward trend resumed after the wash.

The site provided a computer model of the heat and material balance for the plant, which you have used to relate losses in turbine efficiency and flow passing ability to the plant heat rate and output. These relationships are shown in equations (10.1) through (10.7).

The change in steam cycle heat rate due to changes in the HP turbine performance is given by:

$$\Delta HR_{\eta HP} = -HR \cdot \left(0.1825 \cdot \Delta \eta_{HP} \right) \tag{10.1}$$

The change in power output due to changes in the HP turbine efficiency is given by:

$$\Delta P_{\eta HP} = 127.1 \cdot \Delta \eta_{HP} \tag{10.2}$$

A change in steam-cycle heat rate due to the changes in LP turbine efficiency is:

$$\Delta HR_{\eta LP} = -HR \cdot 0.5288 \cdot \Delta \eta_{LP} \tag{10.3}$$

The plant output change due to changes in LP turbine efficiency is:

$$\Delta P_{\eta LP} = 296.96 \, \Delta \eta_{LP} \tag{10.4}$$

Output degradation due to changes in the HP turbine inlet flow function is:

$$\Delta P_{\Phi} = 182 \, \Delta \Phi \tag{10.5}$$

where:

$$\Delta \Phi = -0.144 \left(\frac{t_{effective}}{365} \right) \tag{10.6}$$

$$t_{effective} = t - \left[\left(0.5357 \cdot wash\ cycle\right)\right] \cdot integer\left(\frac{t}{wash\ cycle}\right) \qquad (10.7)$$

and

wash cycle = optimized number of days between water washes,
280 days for the base case

Economic data for the case study is shown in Table 10.4.

10.5.2   Exercises

The exercises below are intended to provide a model for analyzing and calculating an optimum repair cycle period when performance degradation occurs between maintenance cycles. The preventive repair frequency has been modeled with data that yields an optimum cycle identical to the manufacturer's recommendation using the normal distribution function.

Part 1: Setup

1. Determine a curve fit and an equation to represent the data in Table 10.3.
2. Use the result of 1. above to model HP efficiency for the first 1050 days following a major overhaul using a repair cycle of 2 years and water washes of the HP turbine every 280 days.

Table 10.4   Economic data.

Parameter	Value
Plant size (MW)	600
Heat rate after an overhaul (kJ/kWh)	9654
Average capacity factor (%)	92
Delivered fuel cost ($/te)	$35.27
Fuel higher heating value (kJ/kg)	19 920
Nonfuel O&M ($/kWh)	$4.47
Inflation index (%/year)	2.50%
Corporate discount rate (%/year)	10%
Corporate tax rate (%)	28%
Summer/winter power price ($/MWh)	$52.50
Spring/fall power price ($/MWh)	$32.00
Normal major maintenance cost ($)	$4 000 000
Normal major maintenance outage duration (weeks)	6
Differential failure cost	$4 000 000
Differential failure outage duration (weeks)	2
Manufacturer recommended maintenance interval (year)	5
Assumed mean time between failures (year)	6
Assumed failure standard deviation (year)	1

3. Calculate the short-run marginal cost of producing power immediately after a major overhaul and the gross earnings on power sales at the annual average power price neglecting fixed operating and maintenance costs in \$/MWh. (Short-run marginal cost includes only variable operating and maintenance costs.)
4. Using the results of 3. above, determine the short-run lost production earnings due to a normal major maintenance outage in the spring or fall.
5. Using the results of 3. above, determine the differential short-run lost production earnings between normal and failure maintenance with the failure outage occurring in either the summer or winter.
6. Calculate the changes to heat rate and power output after 7.5 months due to changes in the HP, and LP turbine efficiencies, and the inlet flow function.
7. Neglecting inflation, and the discount rate, calculate the short-run marginal cost using the results of 6. above.
8. Using the results of 7. above, calculate the after-tax lost revenue per hour (neglect inflation and discounting).

Part 2: Maintenance Optimization

1. Write a program, or use a mathematics utility, to integrate the costs listed below and determine the net present value over a 30-year life. Use the inflation index and discount rate of Table 10.4.
   (a) Normal major maintenance conducted on schedules between 1 and 6 years.
   (b) The risk of a failure from a normal probability distribution defined by a meant time to failure and standard deviation shown in Table 10.4. See Part 1, number 5. above.
   (c) The costs due to lower efficiency including output and cycle heat rate. See Part 1 number 8. above.
2. Using the program from 1. above, calculate the net present value of the maintenance costs for repair intervals from 1 to 7 years for each half year over the 30-year life of the facility. Complete the following set of calculations for each repair interval:
   (a) Major maintenance cost plus the risk of loss
   (b) Major maintenance plus the risk of loss and efficiency losses.
3. Determine the optimum repair interval for 2 a. and 2 b. above. What does this indicate about the major maintenance program for the facility? How much would the company save by using the optimum repair interval of 2 b. rather than 2 a.?

Part 3: Capital Improvements

Preliminary results of the root-cause analysis team show that copper deposits on the HP turbine blades are the most likely cause for lower turbine efficiency and inlet flow function. They are recommending more water washes to reduce losses and improve earnings. The two teams realize the source of the copper is the condenser and feedwater heater tubes, which were manufactured from a copper alloy material.

1. Repeat Part 2, 2 b. above for a water-wash interval of 90 days.
2. Repeat Part 2: 2 b. above assuming that the copper alloy tubes are replaced with either stainless steel or titanium. The result is a complete elimination of HP turbine deposits, and the inlet flow function remains constant. You expect that the HP

turbine efficiency will fall; but only at 20% of the observed rate. No impact on the LP efficiency is expected.

3. How much could the company spend on condenser, and feedwater heater material replacements to remove copper from the feedwater system? What would be the new optimum major maintenance interval?

4. Create a "waterfall" chart showing the impact on heat rate and output due to HP, and LP turbine efficiencies, and the inlet flow function.

5. List three recommendations for the management committee.

10.6   Closure

The technique developed in this case study can be applied to any piece of equipment or system. It is most applicable for situations where:

- The cost due to performance losses is significant compared to the cost of normal major maintenance, or equipment overhauls.
- There is a long interval between overhauls.
- The timing of overhauls is not stringently dictated by seasonal pricing. Some flexibility should be available.

As shown by the case study, there are instances in which the appropriate maintenance interval should consider equipment performance together with the risks of failure. In some cases, this technique will determine an interval shorter than either CBM or preventive maintenance indicating that more maintenance, not less, is the path to higher returns for investors. It can also be used to provide economic justifications for capital improvements.

Application of this technique is not limited to large complex systems. Replacement of automobile tires is a fine example. It is often prudent and cost effective to replace tires prior to wearing the tread down to the wear bar.

In this case, the engineering analysis of the system revealed that copper used in heat exchanger tubes entered the feedwater and was passed on to the turbine with the steam. A project to replace the heat exchangers with units designed for stainless steel tubes was started immediately. Results of the economic analysis of maintenance practices helped justify the replacement project.

10.7   Symbols and Abbreviations

Btu: British thermal unit
CBM: condition-based maintenance
HP: high-pressure steam-turbine section
HR: heat rate (fuel burned per unit time/power produced)
IP: intermediate pressure steam-turbine section
LP: low pressure steam-turbine section
MM: 1 000 000
NPV: net present value
O&M: operations and maintenance
P: pressure (kPa, psi)
P: power, usually kW or MW

t:	ton, or short ton (2000 lb)
t:	time
te:	metric tonne, or long ton (1000 kg)
therm:	100 000 Btu
T:	temperature (K, R)
v:	specific volume (m³/kg, ft³/lb$_m$)
Δ:	change operator
η:	steam turbine expansion efficiency
Φ:	inlet turbine section flow function $= \dot{m}\sqrt{\dfrac{v}{p}}$ (units of area)

Subscripts
OH: overhaul

10.8   Answer Key

Section 10.4.1: Reading

1. You are probably getting good advice but it would be prudent to have the mechanic show you the tire ware to be sure. According to NASA research, hydroplaning is dependent on a number of parameters including tire inflation pressure, road conditions, vehicle speed, and the depth of water film on the road. As there are a number of parameters out of the driver's control, maintenance of good tire tread is a prudent means of maintaining safe operation of the vehicle.
2. The mechanic has advised tire replacement based on tire performance rather than meeting the minimum safe condition indicated by tread wear-bars. The advice has the objective of optimizing overall operating costs by reducing accidents caused by hydroplaning on wet roads rather than replacing tires based on meeting a specific ware condition.

Section 10.5.2: Exercises

Part 1: setup

1. The author suggests creating a curve fit (Figure 10.5) for the change in HP efficiency expressed as positive numbers. A liner function of the natural log of time or an exponential function work quite well. The author chose the log function. If these types of functions are used, error trapping for zero, and values near zero are required. Water washes effectively translate the time axis a percentage of the elapsed time between washes. In this case, efficiency degradation resumes at a point in time as shown by equation (10.7). In a similar manner, the time axis is also translated back to time = 0 after each overhaul as shown by equation 10.8.

$$t_{OH} = t - \left(repair\,interval\right) \cdot integer\left(\frac{t}{repair\,interval}\right) \tag{10.8}$$

2. The equations above yield Figure 10.6.
3. $21.56/MWh.
4. A 6-week outage with power prices of $32/MWh yields $5.8 million of lost marginal revenue.
5. A 2-week extension of the normal major overhaul outage occurring in the summer or winter yields lost differential marginal revenue of $13.8 million.

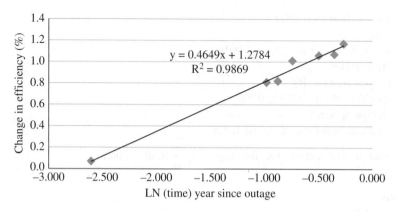

Figure 10.5   Efficiency curve fit option.

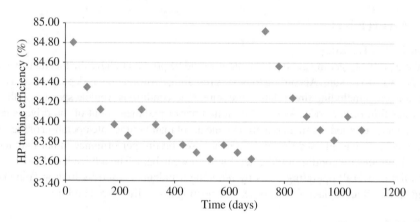

Figure 10.6   High-pressure turbine efficiency model.

6. Changes after 7.5 months:
   (a) $\Delta\eta_{HP} = -1.06\%$;
   (b) $\Delta\Phi = 9\%$;
   (c) $\Delta\eta_{LP} = -1.06\%$;
   (d) $\Delta HR_{\eta HP} = 19\,kJ/kWh$;
   (e) $\Delta P_{\eta HP} = 1.4\,MW$;
   (f) $\Delta HR_{\eta HP} = 54\,kJ/kWh$;
   (g) $\Delta P_{\eta LP} = -3.2\,MW$;
   (h) $\Delta P\Phi = -16.4\,MW$.
7. 21.69 $/MWh
8. $332 $/h at a capacity factor of 92%

Part 2: maintenance optimization:

1. See Figure 10.7.

```
Sheet1 - 1

Private Sub Integrate_Click()
Dim y(1), area(1)

' definitions
' MTBF = mean time between failures
' sigma = standard deviation of MTBF
' CPI = Consumer Price Index - assumed
' drate = company discount rate
' PP_Smmer = normal power price in Summer and Winter
' PP_Wpring = normal power price in Spring and Fall months
' NRepair = normal major maintenanc cost
' NLPO = Normal lost production cost during normal major maintenance
' WRepair = repair cost for major failure
' WLPO = Worst case lost production for major failure

tol = 0.0005
' Obtain constants for problem
Call Readit(size, HR, CF, Cost_MMBtu, var_NonFuel, CPI, drate, PP_Summer, PP_Spring,
taxrate, MTBF, sigma, NRepair, NLPO, WRepair, WLPO)

' look at three improvement cases
' 0: basecase as measured
' 1: Shorten the HP wash cycle to 90 days
' 2: Implement capital improvements to dimish losses by eliminating HP turbine deposits
'      assume HP effiency falls at a rate of
'      assume inlet flow function deterioration stops
'
improvcase = 0
While improvcase < 3

If improvcase = 0 Then
    losscase = 0
Else
    losscase = 1
End If

While losscase <= 1

        icycle = 1 ' Number of years between normal maintenance
        ' Run through cycle periods from 1 to 7 years
        Line = 1

        While icycle <= 7
                ' set flage for maintenance interval checking
                maint = 0 ' Normal maintenance not yet complete for the current cycle
                ncycles = 1 ' current maintenance repair cycle: 1 to project life / icycle

' set initial conditions
x = 0
i = 1
j = 0
k = 0
area(0) = 0
area(1) = 0

While i < 60
    area(k) = 0
    While x <= 30
        If j = 0 Then
                Call f_W(RepairCost, x, icycle, maint, ncycles, MTBF, sigma, CPI, drate,
                NRepair, NLPO, WRepair, WLPO)
            If losscase = 1 Then
                    Call f_eta(improvcase, eta_cost, x, icycle, size, HR, CF, Cost_MMBtu,
                    var_NonFuel, CPI, drate, PP_Summer, PP_Spring)
            End If
            y(j) = (RepairCost + eta_cost) * (1 - taxrate)
            j = 1

            Else
                Call f_W(RepairCost, x, icycle, maint, ncycles, MTBF, sigma, CPI, drate,
                NRepair, NLPO, WRepair, WLPO)
            If losscase = 1 Then
                    Call f_eta(improvcase, eta_cost, x, icycle, size, HR, CF, Cost_MMBtu,
                    var_NonFuel, CPI, drate, PP_Summer, PP_Spring)
            End If
            y(j) = (RepairCost + eta_cost) * (1 - taxrate)
```

Figure 10.7   Example Visual Basic cost integration program.

```
Sheet1 - 2

            End If
            If x > 0 Then
                  area(k) = area(k) + (y(0) + y(1)) / 2 * (1 / (4 * i))
                  y(0) = y(1)
            End If
            x = x + 1 / (4 * i)
        Wend

        If k = 0 Then
            k = 1
        Else
            Delta = Abs((area(1) - area(0)) / area(0))
            area(0) = area(1)
            If Delta < tol Then GoTo LastLine
        End If
          i = i + 1
          x = 1
          j = 0
          y(0) = 0
          y(1) = 0
          maint = 0
          ncycles = 1
      Wend

LastLine:
            If losscase = 0 Then
             Worksheets("Prob9").Cells(6 + Line, 7).Value = icycle
            End If
            Worksheets("Prob9").Cells(6 + Line, 8 + losscase + improvcase).Value = area(1)/1000000

            icycle = icycle + 0.5
            Line = Line + 1
        Wend

    losscase = losscase + 1
Wend
'Worksheets("Prob9").Cells(7, 4).Value = c
'Worksheets("Prob9").Cells(8, 4).Value = i
'    increment the improvement case
improvcase = improvcase + 1
Wend

End Sub
Private Sub f_W(RepairCost, x, cycle, maint, ncycles, MTBF, sigma, CPI, drate, NRepair,
NLPO, WRepair, WLPO)

' Calculates the risk of failure and NPV repair costs for a failure and normal maintenance

' Check to see if it is the right year for a normal overhaul
' For a normal major maintenance year set maintenance flage

If x >= cycle * ncycles And x < cycle * ncycles + 1 Then
      maint = 1
      cNormal = (NRepair + NLPO) * (1 + CPI) ^ Int(x)
End If
' after the end of the major maintenance year, reset maintenance flag and update the counter for
' the current repair cycle

If maint = 1 And x >= cycle * ncycles + 1 Then
      maint = 0
      ncycles = ncycles + 1
End If
' Probability of failure
P = Application.WorksheetFunction.NormDist(x - Int(x / cycle) * cycle, MTBF, sigma, True)
' cost of failure
cFailure = P * (WRepair + WLPO) * (1 + CPI) ^ x

' Calculate NPV

RepairCost = cNormal / (1 + drate) ^ Int(x) + cFailure / (1 + drate) ^ x
```

Figure 10.7   (*Continued*)

```
Sheet1 - 3


End Sub
Private Sub Readit(size, HR, CF, Cost_MMBtu, var_NonFuel, CPI, drate, PP_Summer, PP_Spring,
taxrate, MTBF, sigma, NRepair, NLPO, WRepair, WLPO)

size = Worksheets("Prob9").Cells(3, 2)
HR = Worksheets("Prob9").Cells(4, 2)
CF = Worksheets("Prob9").Cells(5, 2)
Cost_MMBtu = Worksheets("Prob9").Cells(8, 2)
var_NonFuel = Worksheets("Prob9").Cells(9, 2)

' Economic Data
CPI = Worksheets("Prob9").Cells(13, 2)
drate = Worksheets("Prob9").Cells(14, 2)
PP_Summer = Worksheets("Prob9").Cells(15, 2)
PP_Spring = Worksheets("Prob9").Cells(16, 2)
taxrate = Worksheets("Prob9").Cells(18, 2)

' Reliability data
MTBF = Worksheets("Prob9").Cells(21, 2)
sigma = Worksheets("Prob9").Cells(22, 2)

 ' Normal Maintenance
NRepair = Worksheets("Prob9").Cells(25, 2)
NLPO = Worksheets("Prob9").Cells(28, 2)

'Failure Maintenance
WRepair = Worksheets("Prob9").Cells(31, 2)
WLPO = Worksheets("Prob9").Cells(33, 2)

End Sub
Private Sub f_eta(improvcase, eta_cost, x, cycle, size, HR, CF, Cost_MMBtu, var_NonFuel,
CPI, drate, PP_Summer, PP_Spring)

' Calculates the NPV cost due to lower turbine efficiency

' Wash frequency and degradation time improvement index in days
If improvcase > 0 Then
    w_f = 90
Else
    w_f = 280
End If
w_index = (0.5357 * w_f)

' HP turbine efficiency degradation since last major maintenance
a = x - Int(x / cycle) * cycle

Days = a * 365
days_hp = Application.WorksheetFunction.Max(Days - w_index * Int(Days / w_f), 0)

If improvcase <= 1 Then
        ' base case efficiency a power function of time
        If days_hp > 0 Then
                eta_hp = Application.WorksheetFunction.Max(0, 0.012784 + 0.004649 * Log(days_hp/365))
        Else
                eta_hp = 0
        End If
Else
        ' assumed efficiency curve after capital improvements
        If Days > 0 Then
                eta_hp = 0.25 * Application.WorksheetFunction.Max(0, 0.012784 + 0.004649 * Log(Days/365))
        Else
                eta_hp = 0
        End If
End If
' Efficiency Power lost
DeltaPower_hp = eta_hp * 127.1

' Increase in cycle heat rate (Btu/kWh)
deltaHR_hp = HR * 0.1825 * (eta_hp)

If improvcase = 2 Then
    phi_s = 0
```

Figure 10.7   (*Continued*)

```
Sheet1 - 4

Else
     ' Inlet flow coefficient impact on output measured
     phi_s = 0.144
End If
' Inlet flow coefficient impact after capital improvements
' phi_s = 0
days_phi = Days - w_index * Int(Days / w_f)
delta_phi = days_phi * phi_s / 365
DeltaPower_phi = 182 * delta_phi

' LP Efficiency degradation
eta_LP = 0.017 * (x - Int(x / cycle) * cycle)
deltaHR_lp = HR * 0.52876 * eta_LP
deltaPower_LP = 296.96 * eta_LP

' degraded power output
dPower = size - DeltaPower_hp - DeltaPower_phi - deltaPower_LP
'cost_deltaHR = deltaHR_hp * dPower * Cost_MMBtu * 8.76 * CF * (1 + CPI) ^ x / (1 + drate) ^ x

' change in fuel costs
cost_deltaHR = size * (deltaHR_hp + deltaHR_lp) * Cost_MMBtu * 8.76 * CF * (1 + CPI) ^ x / (1 + drate) ^ x

avg_PP = (PP_Summer + PP_Spring) / 2

' Variable cost to produce power

VariableCost = (HR + deltaHR_hp + deltaHR_lp) * Cost_MMBtu / 1000 + var_NonFuel

' Change in revenue

cost_dpower = (DeltaPower_phi + DeltaPower_hp + deltaPower_LP) * (avg_PP - VariableCost)* 8760
* CF * (1 + CPI) ^ x / (1 + drate) ^ x

eta_cost = cost_dpower + cost_deltaHR

End Sub
```

Figure 10.7   (*Continued*)

Table 10.5   Repair cycle optimization.

Cycle period (year)	Base case without losses	Base case with losses
	NPV ($MM)	
1	82.8	103.0
1.5	54.8	84.3
2	40.1	76.2
2.5	31.3	74.9
3	25.5	76.1
3.5	21.8	78.5
4	18.9	82.3
4.5	17.0	86.8
5	16.1	92.9
5.5	17.5	99.6
6	19.8	109.2
6.5	23.8	117.8
7	29.5	129.5

2. See Table 10.5.
3. Without performance losses, optimum = 5 years, with losses 2.5 years. This implies that operating on the manufacturer's recommended schedule is not enough maintenance for this facility. Net present value can be improved by almost $18 million by performing maintenance twice as often.

Part 3: capital improvements:

1. See Table 10.6 and Figure 10.8.
2. Increased washing improves returns but does not change the optimum overhaul interval from 2.5 year of the base case. Capital improvements that remove deposits from the HP turbine increase the optimum cycle to four years, based on the assumptions stated. Replacements of feedwater heaters and the condenser with noncopper alloys could potentially save $30 million.
3. See Figures 10.9 and 10.10.
4. Recommendations may include:
   (a) Begin an immediate change to the overhaul schedule to 2.5 years
   (b) Shorten the water wash cycle to 90 days

Table 10.6 Ninety-day wash cycle and capital improvements.

Cycle period (year)	Ninety-day wash cycle	Capital improvements
	NPV ($MM)	
1	99.3	89.6
1.5	79.3	64.4
2	72.1	53.3
2.5	70.5	48.0
3	71.6	45.1
3.5	74.4	44.3
4	77.8	44.3
4.5	82.4	45.4
5	88.4	47.7
5.5	95.0	51.4
6	104.9	57.2
6.5	113.5	63.4
7	125.1	71.8

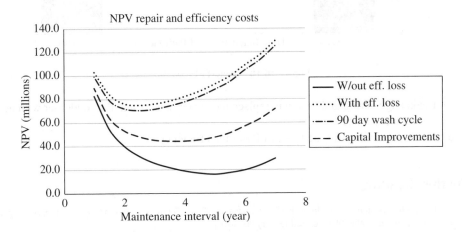

Figure 10.8 Graphic results of repair cycles.

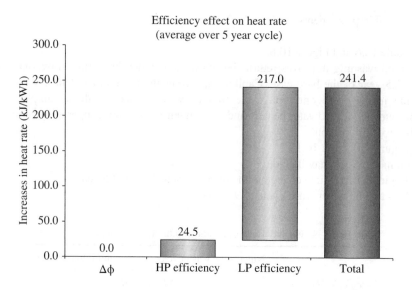

Figure 10.9   Waterfall – heat rate.

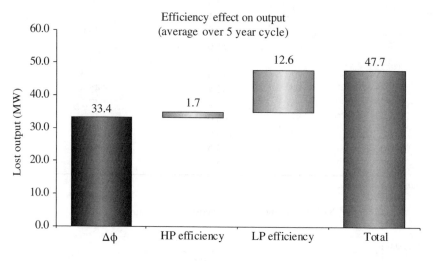

Figure 10.10   Waterfall – output.

(c) Begin engineering and design to replace feedwater heaters and the condenser with non-copper alloy tubes

(d) Investigate causes for the loss in LP turbine efficiency

Further Reading

Berger, D. (2006) Six steps to condition based maintenance. *Plant Services Digital Magazine,* http://www.plant services.com/articles/2006/199/(accessed January 30, 2016).

Fixed and variable expenses (n.d.), http://www.inc.com/encyclopedia/fixed-and-variable-expenses.html (accessed January 30, 2016).

Lutchman, R. (2006) *Optimized Maintenance, A Roadmap to Excellence,* http://www.environmental-expert.com/Files%5C5306%5Carticles%5C13838%5C472.pdf (accessed January 30, 2016).

Measuring tire tread depth with a coin (n.d.) *TireRack.Com,* http://www.tirerack.com/tires/tiretech/techpage.jsp?techid=51 (accessed January 30, 2016).

References

Energy Information Administration (2013) *Capital Costs for Electricity Plants,* http://www.eia.gov/forecasts/capitalcost/(accessed January 30, 2016).

Horne, W. B. and Dreher, R. C. (1963) Phenomena of Pneumatic Tire Hydroplaning, NASA, Washington DC, http://babel.hathitrust.org/cgi/pt?id=uiug.30112106773275;view=1up;seq=19 (accessed January 30, 2016).

Sabol, S. B. (1985) Component efficiency and the optimum repair cycle period. Paper delivered at EPRI Heat Rate Improvement Workshop, San Francisco, CA, October.

Case 11

Project Engineering*

11.1 Opening

Following the conclusion of World War I, the once-warring parties were left with stockpiles of chemical warfare munitions, some weaponized others not. World War II, the Korean Conflict, and the Vietnam War, introduced new chemical warfare agents that added complexity and size to the US, and Euro-Asian stockpiles while leaving the original stores intact. The Chemical Weapons Convention drafted in 1992 outlaws the production, stockpiling, and use of chemical weapons and precursors that may be used to manufacture chemical weapons.

In 1992, the US congress passed a Bill in compliance with the Chemical Weapons Convention, which was signed into law (Public Law 102-484) requiring the US army, the owner of the stockpiled and nonstockpiled chemical weapons in the United States, to destroy all its chemical warfare munitions. The army was prevented from transporting any of the weapons or chemicals; therefore, the destruction had to take place at the storage locations. The public law funded the program and set a time limit to complete the destruction.

The baseline technology selected for destruction was incineration. Incineration was complete, and is today the only approved method to classify chemical warfare munitions, storage containers, or equipment once in contact with a chemical warfare agent as a safe to handle. However, opposition to incineration in the continental United States grew in the years following 1992, particularly in cases where the stockpile was located near a dense population center. Nonstockpiled weapons were found in 49 of 50 states, buried in residential neighborhoods, along roadsides or a railroad routes, in remote areas, or near school yards. Incineration

---

* For a better understanding of the chemical demilitarization technologies, further information is available in National Research Council (1996).

Case Studies in Mechanical Engineering: Decision Making, Thermodynamics, Fluid Mechanics and Heat Transfer, First Edition. Stuart Sabol.
© 2016 John Wiley & Sons, Ltd. Published 2016 by John Wiley & Sons, Ltd.
Companion website: www.wiley.com/go/sabol/mechanical

of the nonstockpile munitions was impractical. In 1994 the National Research Council (NRC), which had overseen the army's program of chemical stockpile destruction since 1987, recommended continued research into alternative low-temperature, low-pressure neutralization technologies, the result of which was a project to chemically neutralize a stockpile of 1621 tons of mustard contained in 1818 ton containers located at the Aberdeen Proving Ground in Maryland, a stockpile located just 30 miles from Baltimore, MD and about 70 miles from Washington DC.

You are an engineer with a top-ten architect/engineering firm. Your company has been contracted by the US army to produce a preliminary engineering design for a full-scale production facility based on research experiments to destroy 1621 tons of mustard ("HD") stored at the Aberdeen Proving Grounds in a campaign lasting less than 18 months. The project manager has selected you to function as the senior project engineer to be located at the field office near the proving ground. Sensing liability risk for such a sensitive project, a senior executive of your company has restricted any contact with the company's chemical engineers regarding the project. You will have to hire any chemical engineering support that will be located at the site office.

During the first few months of the project, the team consists of you and the project manager. The NRC will oversee your progress through quarterly reports and direct communication to resolve questions and concerns. You are assigned a contact at the NRC based on your expertise in engineering and project management, and will correspond directly with the contact, answering all his questions.

Economic circumstances and the unique opportunity to serve your county while making lasting professional contacts with some of the most influential individuals in the county propel you into the project and require you to move your wife and young family across country to Baltimore, MD.

Your work will be scrutinized by highly educated, qualified and capable individuals, by supporters, by those who doubt your capabilities, and by individuals who are vying for your job.

11.2   Background

11.2.1   Mustard

Mustard ($C_4H_8Cl_2S$), sometimes referred to as "mustard gas," is actually a liquid at ambient temperature with a freezing point of 14.45 °C. Distilled mustard (HD) is a blister agent, hazardous upon contact, and a known carcinogen. Due to its low solubility in water it can remain in the environment for a long time and is thus classified as a persistent agent. Even in the distilled form, it contains a number of impurities subject to strict hazardous waste regulations, including 1,2-dichloroethane, tetrachloroethane, and hexachloroethane among other compounds. In addition, the containers of mustard at Aberdeen have developed a solid "heel," consisting mainly of sulfonium, iron salts, and absorbed HD, which readily dissolves in hot water. The quantity and makeup of the heel varies from container to container. The heel consists of small BB-sized particles up to larger chunks, about the size of a fist.

Figure 11.1 shows HD diagrammatically. Material from the Aberdeen containers is dark brown, almost black in color. Other physical and chemical properties are shown in Table 11.1.

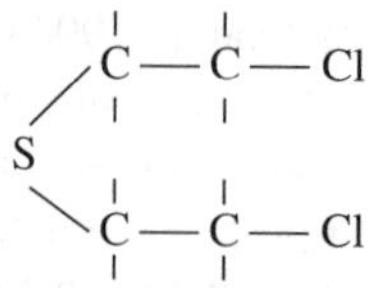

Figure 11.1 HD composition and structure.

Table 11.1 Selected properties of HD.

Characteristic	
Molecular weight	159.08
Boiling point (°C)	217
Freezing point (°C)	14.45
Vapor pressure at 20 °C (mm Hg)	0.072
Volatility (mg/m³)	75 at 0 °C
	610 at 20 °C
Surface tension (dynes/cm)	43.2 at 20 °C
Viscosity (cS)	3.95 at 20 °C
Liquid density at 20 °C (g/cm³)	1.2685
Solubility (g/100 g of distilled water)	0.92 at 22 °C; soluble in acetone, CCl_4, CH_3Cl, tetrachloroethane, ethyl benzoate, ether
Heat of combustion (cal/g)	4.5

Source: National Research Council.

11.2.2 Working with Warfare Agents

Chemical warfare agents are a special class of hazardous chemicals designed or selected for their effectiveness in a particular combat or strategic situation. Agents may be nonlethal – tear gas for example – or lethal, as in the case of nerve and blister agents. Lethal agents may be gaseous or liquid, persistent or nonpersistent. Persistent agents linger in the environment and may be used to hold an area, preventing enemy penetration, whereas use of a nonpersistent agent would allow immediate occupation. Lethal agents differ from other hazardous situations, such as radioactive material, in that there is no lower limit for exposure. A worker handling radioactive material, for example, may safely work until a set exposure limit is reached, whereas the exposure to the slightest quantity of a warfare agent may be fatal.

Unlike nerve agents, there is no antidote for the harmful effects of HD. Therefore, technicians trained to handle agents prefer to avoid working with HD and other blister agents even though the lethality of HD is much less than for nerve agents. Working with live agents is hot, difficult, and strenuous, limiting work times. Working in "Level A" containment – in the presence of a live agent – requires a self-contained breathing apparatus (SCBA), a whole-body garment made of impenetrable material, with sealed openings at the wrist and angle among other requirements. Usually three pair of gloves are worn due to the corrosive nature of the agents and their tendency to pass through synthetic and natural materials. A clean air supply may be provided through a hose from an external source or from SCBA equipment and compressed air tanks worn inside the full-body suit. The suits may be vented through a one-way valve configuration that is part of the suit. Level A requires steel-toed shoes, and two-way communications equipment worn inside the suit.

Working times for Level A work may be limited by the air supply, and could be as short as 15 to 20 minutes for very strenuous conditions but are generally no longer than 2 hours. Each entry into a Level A area requires at least three people. Two enter the area to perform the actual work, and one stands guard at the exit in the event of an accident. After use, the protective equipment is decontaminated to allow undressing. The Level A suit materials may be washed for reuse, or discarded depending on the materials and agents handled.

Decontamination is classified by a number of Xs. Three-X decontamination is generally accomplished by chemical washing, most often an aqueous solution of sodium hydroxide. Three-X material may be removed from a Level A containment and transported. It should not be handled without protective equipment, especially for persistent agents such as HD. Five-X decontamination is accomplished by incineration. After five-X decontamination, equipment that has come in contact with a warfare agent could be declassified as not being a weapon. Only five-X equipment or material can be released from the army's control.

11.2.3   Alternative Technology for HD Decontamination

The army, together with the National Research Council, selected direct hydrolysis of HD with water followed by biodegradation of the hydrolysate as the best alternative to the baseline destruction method – incineration. Though HD has a low solubility in water, the chlorine bond, the essential ingredient to the agent as a weapon, reacts readily in hot water. The hydrolysis reaction products are thiodiglycol and hydrochloric acid. The overall reaction is shown in equation (11.1). The reaction is exothermic, releasing about $15\,kcal$ per mole of HD. To prevent reforming HD and the generation of unwanted side reactions, agent concentration is kept to under 10% by weight, and sodium hydroxide is added after the completion of the hydrolysis reaction to neutralize the acid.

$$S\left(C_2H_4Cl\right)2 + 2H_2O \leftrightarrow S\left(C_2H_4\left(OH\right)\right)2 + 2HCl \tag{11.1}$$

Following hydrolysis and acid neutralization, biodegradation was shown to break down the thiodiglycol successfully to carbon dioxide, water, and H_2SO_4, while generating biological growth. The sulfur acid from the oxidation of thiodiglycol requires neutralization and buffering to maintain a neutral pH for the microorganisms. The biological population is derived from sewage sludge adapted to use thiodiglycol as the only food source. The microorganisms will require additional nutrients to maintain a healthy colony.

11.3   Project Planning and Definition

Engineers are trained in fundamental courses to read and understand a problem, collect any necessary data, do any required research, perform calculations to solve a problem, and turn in their results for a grade. As a lead project engineer, this is not necessarily the right approach. For this project, there will be various development and execution phases, and iterative design process, many different disciplines involved in various aspects of the design, diverse organizations involved with requirements and review of work products, oversight by the National Research Council, equipment vendors and service providers that will need to guarantee products never before used for this specific application, government oversight, stakeholders in the community, researchers, technicians, and others. Therefore, a project-execution strategy needs to be developed in order to complete the required deliverables successfully.

In general, projects can be broken down into several phases. Each phase is conducted as a complete project, beginning with an initiation phase, progressing to planning, execution, control and project close out, all of which overlap in the execution schedule. As the project evolves through each phase, the details become clearer with better definition of the project scope, WBS, schedule, budget, and so forth.

Commercial projects often develop through several development phases such as identification of an opportunity, selection of a single option, optimization, front-end engineering and design (FEED), and then detailed design followed by construction. Between each phase the company developing the project typically conducts a phase gate review to ensure the project economics remain vital, risks are controlled or mitigated, the schedule is realistic and consistent with the economics, project stakeholder input is incorporated, legal requirements are satisfied, the project maintains compliance with environmental regulations, and so forth.

The Alternative Technology program is not a commercial endeavor but there are analogous functional roles and responsibilities. The owner, in this case is the US army. For other countries that have ratified the Chemical Weapons Convention (CWC), the owner would likely be a similar organization. The US Congress provided funding for the project, and had some requirements similar to a bank or lending institution on a commercial project. For example, Congress required the army to engage the Alternative Technologies Panel of the National Research Council to provide oversight similar to a "banker's engineer" on a commercial project. In another country, there would be an analogous body authorizing funds necessary for the project.

The project does not have a commercial objective; however, there is an authorized budget and an expected schedule for completion. Rather than a charter, contract, or project definition authorizing the project manager to apply resources to complete the project, a public law required by ratification of the CWC initiated and authorized funds for the project. The public law specified the high-level project requirements to destroy the stockpiles of chemical weapons. The Alternative Technology program was essentially a change order brought about by stakeholder concerns over baseline incineration technology.

The army's project team is headed by a general who reports directly to the Secretary of Defense appointed by the President. The general is responsible for the interface with the National Research Council, baseline technology operations underway to destroy chemical weapons, and the specific alternative technology HD destruction project headed by a lieutenant colonel. Resources available to the general include the research laboratory and scientists at Aberdeen, MD and elsewhere, as well as test facilities designed for live agents. Your company is a subcontractor to a general contractor working with the army on a portfolio of projects and operating bases related to chemical weapons destruction. The US Congress appropriated funds for the army to use for the alternative technology program. Congress will rely on the National Research Council for assurance that the funds are properly and efficiently deployed for the project, and that the army achieves the objective of the complete destruction of the chemical weapons in the time allotted by public law.

The NRC Alternative Technology panel members are from five universities, two national laboratories (Los Alamos and Sandia), the Bureau of Mines, various consultants and industry experts. Six of the members were on the Stockpile Committee with detailed knowledge and experience dealing with issues of chemical weapons destruction. Eight new members were selected based on their knowledge and expertise to evaluate the alternative technologies thoroughly.

Your company's project manager has elected to execute the project with a "virtual" team. The process design, customer interface, and project control will be at the project office. Civil/structural, electrical, instrumentation and control, facility engineering, architecture, and design expertise will be handled in the home office. Subject matter experts for rotating equipment, pumps, wastewater treatment, construction materials, and so forth, will also be provided on an as-needed basis from the home office. The project manager has hired a consultant with broad experience and chemical processing knowledge who has recently retired from a well know chemical company. Project meetings with the client will generally be with the project manager and you. Other team members may be involved depending on the purpose of the meeting.

Your role on the project will be to lead the conceptual and detailed design of the process – chemical neutralization by hydrolysis followed by biodegradation. You will hire individuals for the process team, ensure timely engineering work products, control the quality of project deliverables, and ensure the company's methods and practices are followed throughout the design process. You will also assist the project manager with communications, inside the company, with the prime contractor, the client – the US army – and communications with your contact at the NRC. You will make technical decisions on methods to accomplish the process and complete the process design. You will be responsible for reporting to the client, and the NRC regarding technical and engineering matters, including progress of work, technical decisions, operations and maintenance strategies, equipment and material selection, equipment vendor qualifications, and assurance that selected equipment will function as required for the process. You are also required to know and understand the client's requirements for the project. You will attend regular update meetings with the client and assist the project manager in conveying the project's status and managing the client's expectations. This project will be divided into the following phases:

- conceptual design;
- FEED;
- engineer, procure construct (EPC).

The campaign to eliminate the stockpile is considered as operations and is not within the scope of the project. The scope of this case study addresses the conceptual design phase including chemical, biological, and mechanical process design. The process conceptual design will include:

- process basis of design (BOD);
- block flow diagrams;
- process description;
- process flow diagrams (PFDs);
- piping and instrument diagrams (P&IDs);

- cause-and-effect diagrams (CEDs);
- construction materials for the process equipment;
- material balances;
- energy balances;
- utility requirements;
- control philosophy;
- equipment redundancy requirements for reliability;
- equipment verification studies;
- budget and schedule for the next project phase.

11.3.1   Project Management

Project management is an iterative process. Where it starts, therefore, is somewhat arbitrary. In the beginning, the project plan shows only basic functions. As the project information is developed, greater detail is added to the project plan and changes throughout the project are continually fed back into the plan to assure proper execution, management, and control of the project functions. The case study is related to the process design, so a logical starting point would be to prepare a block flow diagram of the process.

Exercise 1

Given the foregoing description of the alternative technology, sketch a block flow diagram of the chemical neutralization process. Each block represents a major step in the process. Show the main process flow and the major supporting chemical process functions or systems that are necessary. At this point in the development do not include utilities such as electricity or steam generation. Water is an important ingredient, so any process step necessary for the chemical reaction involving water should be included.

Give some thought to how the process will be conducted – in a batch sequence or continuously. Consider how destruction of the agent will be verified. A process step of verification that requires equipment should be included as a block on the block flow diagram.

11.3.2   Client Requirements

The army requires the destruction process to be complete to better than 99.9999% –six nines – which is the given definition of destruction required by the public law. The ton containers are to be decontaminated to 3-X – no detection of agent in the air around the items or on the surface. All the equipment selected for the process is to be verified for functionality. Neutralization reaction conditions are shown in Table 11.2.

Ton container cleanout requirements are shown in Table 11.3.

Parameters governing the biological reactor are shown in Table 11.4.

The biological colony requires nutrients including nitrogen, potassium, phosphorous, and magnesium, to sustain the population. These are provided via feeds of aqueous ammonia, phosphoric acid, potassium chloride and the salts shown below:

nutrient salts;
magnesium sulfate – $MgSO_4 \cdot 7H_2O$;
iron sulfate – $FeSO_4 \cdot 7H_2O$;
calcium chloride – $CaCl_2 \cdot 2H_2O$

Table 11.2   HD neutralization rates.

Parameter	
Agent decontamination	
Destruction rate (lb/day)	8000
Operations	8 hours/day, 5 days per week
Reaction temperature (°C)*	90
Weight faction HD/water	0.04
Reaction time*	20 minutes to a detection limit of 200 ppb weight – allow 1 hour in full-scale system
Other	Vigorous mixing, reactor venting closed after water addition, maximum reactor pressure 1 atm
Acid neutralization	8.5% excess NaOH
Caustic	18% aqueous solution
Terminating pH	12

*Source: National Research Council.

Table 11.3   Ton container data and cleanout requirements.

Parameter	
Total contents (lb)	1783 lb
Heel	Nominal 17%
HD purity (average)	90.2%
Cleanout water rinse	272 kg at 90 °C
Steam spray	234 kg/1000 kg HD (860 kPa saturated)

Source: National Research Council.

Table 11.4   Biodegradation parameters.

Parameter	
Reactor type	Sequencing batch reactor
Feed concentration (gram total organic Carbon/g mixed	Min 0.08
liquid suspended solids	Max 0.1
Air (kg/100 kg HD)	62 295
Aeration sequence (hr/day)	17
Microbe generation (g dry solids/g TOC)	0.8
Hydraulic retention time (days)	10
Solids retention time (days)	15
Thiodiglycol destruction	99%

Source: National Research Council.

Following biodegradation of thiodiglycol, the reactor will be decanted and sent to a wastewater treatment works for stabilization of the colony by aerobic digestion. The digested colony will be filtered to remove the biosludge, pressed to a cake, and sent to a landfill. The water from the bioreaction will be discharged to the environment.

Maintenance of equipment that has been in contact with the chemical agent is to be minimized. In most cases, equipment requiring maintenance is decontaminated to 3-X, disposed, and replaced with a new item. Maintenance must be performed with a worker's hands in clear view. Workers cannot, for example, reach behind something to perform maintenance, pick up, or drop an item. Whenever possible, an item that has been in contact with the agent will be handled indirectly, with another tool. A bolt that is wet with agent, for example, will be picked up with a pair of pliers to avoid wetting the worker's gloves.

Operations are to be automated or remotely controlled to the maximum extent possible to reduce entering Level A areas. The stockpile is to be decontaminated in 18 months without transporting the material from its vicinity. At the completion of the campaign, the destruction facility will be dismantled so that it cannot be used for any other purpose. Work is to take place only in daylight hours. As shown in Table 11.3, operations will be one 8-hour shift per day, 5 days per week.

After understanding the client's requirements a little better, return to the block flow diagram and make any changes necessary.

11.3.3   Work Breakdown Structure

Organizing the deliverables required for the conceptual design into a work breakdown structure (WBS) is a planning step in every project that feeds into the development of the project schedule, cost estimate, human resource planning, and acquisition of resources. A WBS segregates the project scope into manageable elements by deliverable. Dividing and subdividing the scope allows for control and management of project though the completion of each project phase to ensure deliverables are completed on time, within budget and that the scope is completely covered for the client. Changes that occur during the project can impact the WBS, so the WBS is often reviewed and updated as necessary throughout the project.

A standard WBS may be available for commercial projects typically executed by your company. It is depicted as a hierarchy similar to a project organization chart. The WBS is organized by function – engineering, project controls, purchasing, legal, and so forth. Each functional area is then subdivided again and again until the scope of the project is adequately defined for the specific project.

Exercise 2

Review the block flow diagram and the scope of the conceptual design. Develop a high-level WBS for your project scope.

After developing the WBS, each element can be broken down into individual activities – things to do. Agent neutralization P&IDs, for example, may be broken into calculations of line sizes and pressure losses, calculation of vessel elevations and pumping requirements, drafting, checking, and revising each drawing.

Check the block flow diagram for any changes resulting from the WBS.

11.3.4   Growing the Team

Candidate Search

Corporate resources for drafting, and mechanical systems, will be provided to the project from existing staff. Two individuals will relocate to the project office for the duration of the first two project phases. Resources for chemical engineering and wastewater treatment engineering will be new hires.

Exercise 3

Write a job description for a chemical engineer whom you will hire. Review the block flow diagram, scope of the project, and the WBS to define the skills the individual must possess. Do not mention the project but do mention the project location, your company, and the basic functions or deliverables for which the individual is responsible.

The human resources department runs an advertisement in the local newspaper. It is posted in an online job-search engine, and on your company's LinkedIn site. After 3 weeks, the human resources department sends you a package containing approximately 100 applications with education, job histories, and individual resumes.

A disadvantage of hiring a new employee specifically for the project is that the staff will be out of a job at the close of the project. Towards the end of the project you will likely not have their undivided attention as they will be looking for employment. Occasionally, your corporation will keep the employee at the end of the project but it could mean relocation, which could be an issue for the individual. Throughout the process of hiring and execution of the project you need to keep the individual's interest in mind. Holding regular discussions related to continued employment, and taking actions that would assist the candidate can help maintain productivity and engagement in the project.

Interviews

Before you get to the interviews, you must reduce the stack of potential candidates to three or four. You realize this is a difficult business – a large number of people will be turned away without a job. But your current task is to select the right person who will report to you, and upon whom you must rely to complete a significant portion of the WBS.

Exercise 4

Write down five criteria to help you sort through the list of candidates and select the top four candidates who will be called in for an interview.

There are a number of interview techniques. Startup IT firms may have one method. Google has theirs. There are at least two reasons for an interview: (i) confirm the client's background and capabilities, and (ii) determine if the individual will work well in your organization. The resume and application go a long way toward accomplishing the first objective. You may have a couple of specific questions that will help confirm if the candidate is properly suited.

The second objective is nontechnical, and requires that you understand how the selected individuals work and behave. Will they be able to work independently? Are they creative thinkers? How do they solve problems and how do they perform under stress? An important aspect of this project is the oversight from the prime contractor, the client, and the NRC, not to mention internal quality control of products and deliverables. There may be aspects of your company's culture, conscience building as opposed to top-down decision making, the review process, or how individuals contribute to project development, for example, that you may also like to explore.

Again there are a number of methods to accomplish this task. One method is to have the candidates explain what they did in a specific situation, and the result of their actions. The interviewer poses open-ended questions on a topic. The candidates need to describe the situation, what they did, and the result. Hypothetical cases are not applicable and should be avoided. You need to hear about a specific case. Also, being "responsible" for something does not mean that any action was actually completed. The candidate must tell you what they did and what resulted.

Open-ended questions are those that typically begin with "what," "how," or "when." There are optional constructions of these types of questions, "Can you tell me a time when. ..," for example. Questions requiring "yes" or "no" answers should be avoided. Leading questions that imply what you are looking for should also be avoided.

You can start this segment of the conversation with a question like: "Tell me about a time when you were faced with a difficult problem." After which, you can dive deeper into the situation by asking leading questions about what the problem was, what they did, how they convinced their supervisor of the solutions, how it helped the company, and so forth.

Use the project manager as a resource. Bounce ideas for questions off the manager and involve him/her in the interview process. After the interviews are complete, compare notes, including impressions of the individuals, how well they responded to the questions, and which candidate should be selected. Ensure you have a consensus with your manager. Your decision is the final word but the manager's experience should be used in reaching the decision.

Exercise 5

Write down two open-ended questions that will help you determine which candidate to select. The questions may be technical or behavioral in nature. They should ensure that the candidates respond with a situation, what they did, and what the results were.

11.3.5 Process Basis of Design

The basis of design (BOD) sets out the project requirements, the conditions of the site, and project assumptions; it specifies the codes and standards to be followed, and the criteria for the success or testing for the project. It can include a process description, equipment lists, and equipment redundancy requirements. Many companies have a standard template for the BOD. It should be drafted, reviewed, and approved by different individuals with appropriate knowledge and authority to perform the functions. As the project progresses and details of the project are refined, the project BOD should be updated accordingly.

The unique nature of this project requires the BOD to include the physical and transport properties of HD and thiodiglycol. Those properties that are a strong function of temperature – viscosity for example – should have more than one point listed in the BOD.

The system of units used by the project must also be determined. The scientific community, with few exceptions, uses SI units with temperature in degrees centigrade and energy expressed in kilocalories. The workers at the completed facility who will operate and maintain the process and equipment will drive about 20 to 25 miles to work and know that a warm day is in the 80s. You will determine the system of units for the project. For a case study considering a stockpile outside the United States the decision may be different.

In addition to the system of units, you will need to provide the client with guidance on the use of Category M piping requirements. Category M substances are defined in the ASME piping code B31.3 as those that will cause significant and irreversible harm from a small exposure, even when prompt restorative actions are taken. The owner, and not the designer, determines the use of Category M piping code but the client has asked your company for guidance and a recommendation.

Exercise 6
1. Research the site and find as many as possible of the following:
    (a) Site latitude and longitude.
    (b) Average and extreme ambient temperatures.
    (c) Maximum and minimum ambient temperatures for the design (temperatures at which the process must function at full capacity without compromising quality).
    (d) Heating and cooling design temperatures.
    (e) Maximum wind speed.
    (f) Prevailing wind direction (seasonal if appropriate).
    (g) Seismic zone classification.
    (h) Average annual rainfall.
    (i) Maximum rainfall event.
    (j) Snow load.
2. Determine the system of units to be used by the project. Write a short justification for the decision.
3. Research Category M piping on the Internet and write a recommendation for the client to use or not to use the classification. If you have access to ASME B31.3 in your company or at a local library, read the definitions and work through the decision tree on Category M. If you recommend using Category M piping, define which systems will carry the designation.

11.4   Executing the Project

The planning phase of the project is almost complete, your team is assembled and everyone is getting to know one another. You selected two highly qualified individuals to be the chemical process engineer and waste treatment engineer. Each has started working on their respective design deliverables and has independently engaged with the client to know and understand the process.

The Process Engineer has conducted live agent experiments at the laboratory with the chief scientist to characterize the flow patterns of agent/water mixtures in piping. The experiments are groundbreaking and necessary for the success of the project. The two immiscible liquids behave much like a vapor liquid system. The dense, viscous HD behaves as a liquid in a vapor/liquid system and forms an annulus on the pipe wall while the lighter, low-viscosity water is confined to the center. Even at high velocities, the two fluids do not mix.

The waste-treatment engineer has shown expertise and demonstrated the need for aerobic digestion of the microbiological colony for the project manager with actual material generated by the army's research team. Treatment studies necessary to demonstrate the feasibility and effectiveness of equipment have relied on this individual's knowledge of vendors, available equipment, and processes.

11.4.1   The Process

The chemical engineer has suggested the following process for HD neutralization:

1. A steam heater heats water to the reaction temperature on its way to a batch reactor.
2. Once filled with water the reactor vents are closed.
3. The reactor will be equipped with a mechanical agitator to provide thorough mixing of water and HD.
4. A recirculation loop draws water from the bottom of the reactor. A pump delivers the water through a static mixer and through spray nozzles located at the top and inside the reactor.
5. The operator verifies that recirculation and agitation are established in the reactor.
6. Ton container clean-out fluid is added to the reactor.
7. HD, pumped from a storage tank, enters the recirculation loop upstream of the static mixer at a weight ratio of 4% over a period of 20 minutes.
8. One hour after the start of HD pumping, the pH provides an indication of a complete reaction.
9. Given a positive indication of HD destruction, 18% aqueous NaOH is added to the reactor to bring the pH to 12 over a period of 10 minutes.

11.4.2   Stakeholder Communication

Early in project planning, the project manager assigned you a contact at the NRC. You are to respond directly to the stakeholder in response to his questions. Besides his work on the NRC, he is also the chemical engineering department head at a prestigious university. In addition to answering questions, you are instructed to form a relationship with your contact to avoid any confusion throughout the project development and execution.

On Friday afternoon, the project manager sends your first question from the NRC. They would like have a reaction mass balance for the neutralization of HD on Monday. The NRC will use your balance in their preliminary report to Congress due at the end of the month.

Assume there are 12 neutralization batches per day to accomplish the client's objective of disposing of 4 tons of HD each operating day. The proportional quantity of heel and ton container cleanout water and steam are included in each batch.

Exercise 7

Determine the mass balance for the destruction of HD for a two-step process: agent decontamination followed by acid neutralization. Update the block flow diagram if required. Write an e-mail to the NRC contact with your results.

Be aware when completing Exercise 7 that the project is at a very early stage of development. The client has not vetted the process, has not reviewed any drawings, and has not approved any of the deliverables. Information thus far provided is subject to change. Providing a level of detail that implies the process has been vetted and approved could potentially trap the project into a procedure that would be difficult to change later in the design process. You therefore need to develop a presentation format that provides sufficient accuracy for an engineering design without the implication that the process procedure is fixed. State any assumptions necessary.

Exercise 8

Your NRC contact appreciated the results but would like to know the final reactor temperature. Assume the ton container cleanout effluent temperature is 25 °C. To prevent freezing, the ton containers will be heated to 5 °C above the normal freezing temperature during the winter. During the summer months the ton containers will be as the normal average temperature. There are several values for the specific heat of HD. Assume you have been provided 0.36 kcal/kg°C. Assume the specific heat of thiodiglycol is the same as HD. Determine the worst-case (highest) temperature for the water to the HD neutralization reactor such that the final chemical neutralization temperature is 90 °C. Calculate the final temperature from the reactor after neutralization of acid at the annual average ambient temperature.

After completing Exercise 8, review the block flow diagram and make any changes suggested by the results.

11.4.3   Ton Container Cleanout

A disagreement has developed within the team regarding how to breach the ton containers and extract the HD. There is clear agreement that using the fill/drain lines through the original valves on the containers would be futile. The solid heel within the containers and the age of the valves will render them useless. One faction of the team suggests using a large band saw to cut the containers in half. The other insists that the containers should be punched with a hydraulic punch and then drained through two holes approximately 5 cm in diameter. The discussions have taken place for a little over a week and both parties are now entrenched with little sign of a compromise or consensus.

Exercise 9

You have a few options available to resolve the conflict, including:

- Have the parties withdraw from the discussions to provide a cooling-off period prior to re-engaging.
- Call a meeting to brainstorm new ideas allowing all team members to participate with the intent of arriving at a consensus.
- Decide for yourself the answer and make a decision on the correct path forward.
- Call in an expert to provide an alternative which can be discussed.
- Call a collaboration meeting to allow a win-win solution.

Weight the options provided above and determine the best path forward.

11.4.4   Demonstration Tests

In most engineering applications, equipment is selected based on characteristic performance metrics, the pumping head at a given volumetric flow rate for a centrifugal pump, or the flow rate per cross-sectional area of a filter, for example. The design experience of the manufacturer combined with the characteristics of the flow stream and performance necessary to fulfill the project objectives are generally enough to specify the functional requirements for an equipment purchase.

In this case, very few pieces of equipment have functioned at the required operating scale with the specific fluids produced by the neutralization of HD. Filter manufacturers can be confident that their equipment will function as required but they cannot answer with positive certainty that it will. Given a potentially lethal situation that could develop from an accident or equipment failure, absolute certainty that equipment will function properly is highly sought.

Therefore, several bench-scale studies are required to substantiate certain technologies with substances produced in the laboratory. These studies target several objectives, including:

- determination of the optimum design parameters to yield a specified product quality, when specified by the client or an environmental permit;
- the product quality for a standard design, when the product quality is unknown;
- maintenance requirements, and durations between start-of-run and end-of-run for cleaning, and so forth;
- inlet and outlet process condition required to achieve the product quality and proper equipment operation;
- utility requirements for proper operation.

Exercise 10
Following biodegradation of the HD hydrolysate and aerobic digestion of the bioreactor decant, operations periodically settles the digester and withdraws a portion of the liquid and biosolids. This flow is filtered and the solids are pressed into a cake and sent to a landfill. The liquid, mostly water and dissolved salts, is to be discharged to the environment. This process has been executed many times at wastewater treatment works but never with the specific colony grown for this project. Write a short statement of work for a filter vendor to test and size a full-scale filter press. Assume the filter must remove 99% of the total suspended solids.

11.4.5 Materials of Construction

Neutral to slightly basic pH fluids can generally flow though carbon-steel piping and components. The HD neutralization reaction, however, yields a dilute hydrochloric acid solution with a very low pH. There are few materials that can handle this aggressive substance and the reliability of the overall system may be compromised if the correct materials are not selected.

Piping and even pumps can be lined with man-made materials resistant to acids. You have seen, first hand, a prototype facility that used glass-lined reactors with glass-lined agitators for the destruction of nerve agents. You are therefore confident that most of the elements can be constructed of materials that provide exceptional reliability for the life of the facility.

However, the reactor for this design will include a recirculation loop with a static mixer. Experiments conducted at the army's laboratory showed that the mustard and water do not readily mix in a pipeline. The process engineer has specified a static mixer in the recirculation line to ensure that the mustard is dispersed adequately in the hot water. The static mixer cannot be lined and must be constructed of a metal.

The materials specialist on the project has suggested either titanium or zirconium. Zirconium, while an excellent choice, has a connotation of high cost with the client, and the project manager has forbidden the use of the "z" word.

A test must be conducted to verify that titanium is an appropriate material. A vendor has been identified that can supply a static mixer to the process engineer's specifications but the

NRC has asked for a verification that the material will function reliably in the application. A test must be conducted to assure the NRC that titanium is an appropriate material. If no metallic components can function reliably in the HD hydrolysate, the engineering team could be faced with returning to the drawing board for a process design.

Exercise 11

The laboratory will produce a quantity of hydrolysate from HD neutralization for an experiment to test material suitability. The materials specialist has chosen to test titanium, zirconium, and a nickel alloy for suitability. What else is required for the experiment? What material would you suggest for this purpose?

11.4.6   Unexpected Events

The army obtained bids for the detailed design, engineering, procurement, and construction of an HD chemical demilitarization facility for the stockpile at Aberdeen, using your project's process design package as the basis. The civil design package was delivered to the site to start construction. Much of the equipment had been ordered and many of the major subcontracts were out for firm quotation. It was at about this time that the terrorist attacks on the two towers of the World Trade Center in New York and the Pentagon in Washington DC occurred. A fourth plane was thwarted and crashed in a field in Pennsylvania. In all nearly 3000 souls were lost.

The Secretary of the Army assembles a team of experts to determine if there are other potential targets and the stockpile of HD at Aberdeen ranks high on the list. The Secretary orders the project to complete HD neutralization in half the scheduled time.

Exercise 12

You are no longer on the project, but the current contractor knows your name from your years of involvement and numerous decisions, and sees you as an expert. You are asked to provide an opinion as to how the campaign can be accomplished in half the time even while much of the key equipment has already been ordered. What would you suggest?

11.5   Closure

This project was a turning point in my career. It is one that I will always remember. I took a position that many would turn down but was rewarded with a leadership position, lifetime contacts, and not just a few stories.

The last ton container of HD at Aberdeen was neutralized in 2006. The project employed a batch process for agent neutralization followed by a hold, test, and release step prior to the discharge of hydrolysis from the Level A containment area. This step ensured the agent was neutralized and provided reprocessing if necessary. The ton containers were punched and drained, then cleaned with a high-pressure water spray inside the containers to remove the heel. Following the water spray, they were steam cleaned to render the containers to a 3-X level for transportation and incineration.

The terrorist attack of September 11, 2001 heightened the focus on the destruction of chemical agents with a resulting shortened campaign. Continuous around-the-clock processing of the agent was permitted by outsourcing the biodegradation to a nearby commercial entity. Hydrolysate was trucked to the facility with a large capacity.

This was a project in which the process design team was headed by a mechanical engineer. This is not unheard of, but it places requirements on the lead to be familiar with another engineering discipline. If faced with this type of situation, rely on the tools, processes and procedures followed by your company to help organize, plan and execute the project. For this project, the department manual included a standard work breakdown structure that was adaptable to this unique project. The manual also contained design standards, methods of work, forms, and so forth, that allowed for an efficient transition for new employees and existing employees who had not worked on process-related projects.

If there are no corporate standards, there are courses and professional societies that can offer assistance. I can suggest a four-day course for the Project Management Professional (PMP) certification and membership in PMI.org. The course material will provide methods to manage a generic project. Additionally, there are numerous publications, including the *Project Management Book of Knowledge* (PMBOK®) and other workbooks that can help develop a working structure for project leadership.

This project benefited from a qualified computer code for much of the fluid systems work that got us through an important quality audit. If we had used a spreadsheet for the calculations, I am confident we would have had a much more difficult time with deliverables and the overall design package. It was also possible for the prime contractor to terminate our contract if we were unable to correct findings from the audit. Most commercial projects can be completed using spreadsheet calculations. However, the protection features of most spreadsheets are easily overwhelmed, resulting in software that can be corrupted with errors and inaccurate changes to formulas quickly propagating throughout the corporation. Engineers and managers have to be especially diligent in checking the project spreadsheets to ensure accurate calculation results. For sensitive projects – those in the nuclear industry for example use only software packages that are qualified for use by an appropriate agency or your corporation.

In addition to the various tools that helped organize the work, the team was made up of exceptional engineers and managers, without whom the project would have been impossible. I am indebted to their expertise and understanding.

Throughout my career, I have met more than a few people who either knew of the project, were impacted by it, worked there, or knew someone that worked at the site or on the project. Not every project has such far-reaching consequences or such a broad array of stakeholders. However there are stakeholders on every project – and keeping them properly informed, and answering their questions with honesty and integrity is as important as completing design calculations, finalizing drawings and specifications, or reporting to your supervisor.

11.6   Answer Key

Exercise 1: Block Flow Diagram

One possible block flow diagram is shown in Figure 11.2. The case study description left it up to the student to specify whether a continuous or batch process would be followed. However, knowing the lethal nature of the fluids and learning of the client's requirement to destroy the agent to less than 200   ppm by weight, a batch process is implied. The block flow should be a high-level overview of the basic steps. During the development of the conceptual design, greater detail will be incorporated in process flow diagrams and later piping and instrument diagrams.

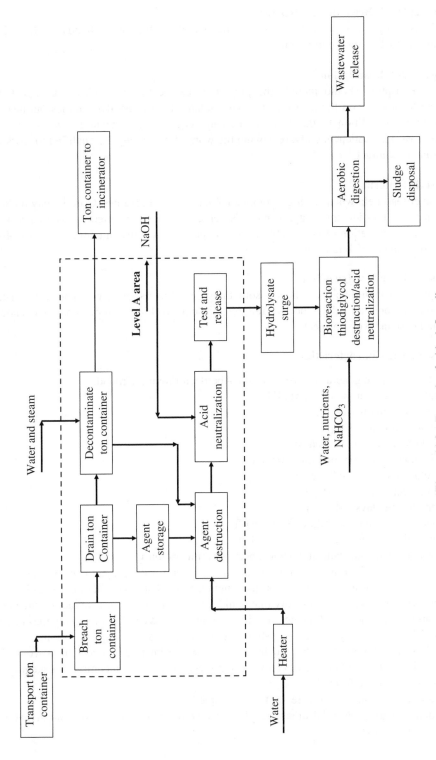

Figure 11.2 Example of a block flow diagram.

Exercise 2: Work Breakdown Structure

The WBS is an organization chart of the project deliverables, which are outlined in section 11.3. An example WBS is shown in Figure 11.3.

Exercise 3: Job Description

The job description should include the project office location and what you expect the candidate to produce on the project. This will include the process deliverables outlined in section 11.3. In addition to the hard engineering requirements, the individual should be capable of working independently, reviewing the work of others, and be qualified to work on a government contract.

Exercise 4: Criteria

Selection criteria help when sorting through a large number of resumes. Criteria may include attendance at a top-tier university, a master's degree, years of experience, experience in a specific area, completion of deliverables, correct spelling, hobbies and community involvement, and willingness to relocate.

Exercise 5: Interview Questions

These questions should be open ended – something a yes/no response would not cover. They can be on any topic including success, failures, difficult working relationships, independent work, and best and worst projects.

Exercise 6: Site Characteristics

1. The following are generally correct for the Aberdeen Proving Ground:
   average ambient temperature: 56.3 °F;
   extreme maximum: 100 °F;
   extreme minimum: 3 °F;
   annual precipitation: 49 inches;
   maximum one-day precipitation: 6 inches;
   maximum snow depth: 9 inches;
   cooling degree days: 1261 (°F);
   heating degree days: 4467 (°F);
   normal maximum temperature: 88.1 (°F) July;
   normal minimum temperature: 26.2 °F January.
   (*Source*: National Oceanic and Atmospheric Administration.)
2. System of units. Traditional English units were chosen for the project because they were familiar to the operating staff. The project engineers felt that unfamiliar units could create an unnecessary level of complexity that could impact actions in tense situations such as a response to an accident, while the scientific community was capable of making the conversions. The student may select any system of units with a reasonable justification.
3. Category M. This exercise has no incorrect answer. Either recommendation would be appropriate with a suitable justification. The project chose not to recommend Category M piping based on the following:
   The areas containing agent were not to be occupied.
   Personnel entering a Level A area would have protective suits with the expectation that live agent was present.

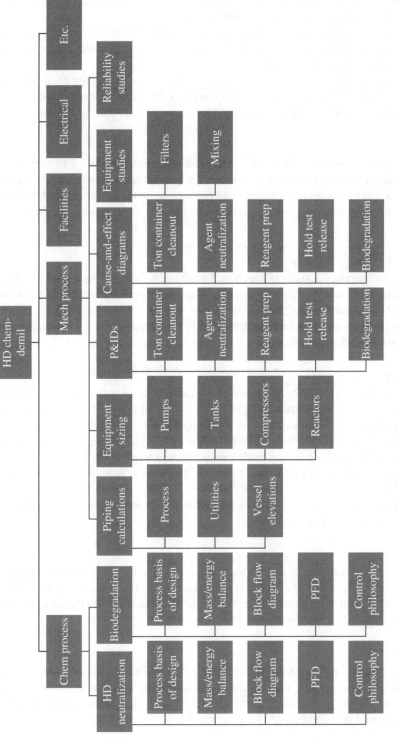

Figure 11.3 Example work breakdown structure.

Operating personnel were trained and had many years' experience working with warfare agents without a Category M classification.

Category M piping requirements include many protective measures that reduce or eliminate the potential for leaks. Prudent application of the standards can reduce corrective actions in the event of a leak.

A recommendation to follow Category M requirements, if elected by a student, should include the limits on where it would be applied. The code requirements can significantly impact costs so the application should be minimized as much as possible.

Exercise 7

Basic results are shown in Table 11.5. The students should propose a format that is easily understood for the presentation to Congress. The project presented the information per unit of HD. The per unit basis allowed the main process variables to be modified without changing data provided to Congress. The results assume that the ton container heel is the same as HD.

Exercise 8: Energy Balance

This exercise requires outside research to determine the specific heat for HCl, and NaOH. Approximate values are sufficient as the overall mixture-specific heat is largely determined by the quantity of water used for the reaction.

Results: initial water temperature: 199 °F, 93 °C; final temperature following acid neutralization: 180 °F, 82 °C.

Exercise 9

The best alternative in this situation would be to make a decision and proceed with the project. The parties have become entrenched and are no longer communicating effectively. The bandsaw option presents a number of problems for operations and maintenance that would place personnel in dangerous situations. The blade can get caught or jammed in the ton container; the blade can break; the spinning blade can distribute a lethal agent throughout the band-saw components. Replacing a sharp blade might cut the Level A suit, exposing maintenance personnel to HD.

The clear choice in this case is to use the hydraulic ram with a blunt point to punch holes in the container and drain the agent.

Exercise 10

The statement of work should include a brief description of the biodegradation process and the requirements of the filter. The objectives of the test should be introduced, then elaborated in some detail so that the vendor is clear about what is required. Elements in section 11.4.4 should be included as appropriate.

Exercise 11

A control group is missing from the outlined protocol. In the case of a study to determine the resistance to corrosion, the control group should be one that is known to corrode in the presence of the hydrolysate. Since 316 stainless steel is known to pit in the presence of HCl, it would be a good choice for the control.

Table 11.5 Material balance results.

	English		SI
Batches per day	12		12
HD and heel processed (ton/day)	4	kg/day	3629
Stockpile (ton)	1621		
Number of ton containers	1818		
Inputs			
Agent and heel			
HD and heel per TC (lb)	1783	kg/ton	809
HD + heel processed (lb/day)	8000	kg/day	3629
HD per batch (lb)	553	kg/batch	251
HD moles per batch (lb moles)	3.48	kg moles	1.58
Heel per batch (lb/batch)	113	kg/batch	51
Heel moles per batch (lb/moles)	0.712	kg moles	0.323
Water			
Total water required (lb/batch)	16 000	kg/batch	7257
Total water moles/batch (lb moles)	888	kg moles	403
Water and steam to TCC (lb/batch)	350	kg/batch	159
Water to reactor (lb/batch)	15 650	kg/batch	7099
Total mass of inputs (lb/batch)	16 667	kg/batch	7560
Agent decontamination outputs			
Thiodiglycol per batch (lb/batch)	512	kg/batch	232
HCL per batch (lb/batch)	306	kg/batch	139
Water remaining (lb/batch)	15 849	kg/batch	7189
Total mass of outputs (lb/batch)	16 667	kg/batch	7560
Acid neutralization inputs			
NaOH per batch (lb/batch)	335	kg/batch	152
NaOH moles per batch (lb moles)	8.38	kg moles	3.80
Water with NaOH (lb/batch)	1527	kg/batch	693
Total mass of inputs (lb/batch)	18 529	kg/batch	8405
Acid neutralization outputs			
Thiodiglycol (lb/batch)	512	kg/batch	232
NaCl (lb/batch)	490	kg/batch	222
Water (lb/batch)	17 527	kg/batch	7950
Total mass of outputs (lb/batch)	18 529	kg/batch	8405

Exercise 12

A response to the current contractor will be an opinion of your company, even though you have been asked directly by an individual on the project. Therefore, any response should be vetted by your company prior to its delivery. From the project data on the reaction rates, you know that the agent neutralization will take approximately 20 minutes while the biodegradation will take 7 days. You also know that the agent neutralization was originally planned to be conducted 8 hours per day, and 5 days per week. Increasing the agent neutralization processing rate would, therefore, easily overwhelm the bioreactors or require several times the bioreactor volume. Therefore, finding a third party or several parties to perform the biodegradation is a reasonable course of action, and the hydrolysate can be shipped as there is no indication of the presence of agent.

Reference

National Research Council (1996) *Review and Evaluation of Alternative Chemical Disposal Technologies,* National Academy Press, Washington DC.

Case 12

In the Woodshop

As an amateur at woodworking, your skill level is steadily improving. You have made a few pieces you are proud of: a Philadelphia low-boy with cabriole legs and carved feet shown in Figure 12.1, a Sheraton table from the "Master Class" section of *Fine Woodworking,* and a modern slant front desk for your spouse. You have not graduated to chairs or designing furniture of your own, but the pieces you completed are well built, and you have become proficient using the tools in your shop, planning projects, and have even gained skill with handcut dovetails.

You have searched for plans for one piece – a reproduction, eighteenth-century, Pennsylvania-style secretary with a bookcase. There are plans for slant-front secretaries available on the Internet and in a few books but they tend to be either too complicated for your skill level, not very appealing, or incomplete. In February, *Fine Woodworking* published the first installment of three Lonnie Byrd articles with complete plans for the Pennsylvania secretary. Beautifully proportioned, lots of instruction, easy to understand plans, and you trust *Fine Woodworking* to publish accurate, tested, and complete plans.

After 6 months, you are completing the finishing touches on the book case. The main secretary base is complete. The tombstone doors were a challenge but, using the instructions, the panels came out perfectly, and the doors open and close smoothly. Reproduction hardware was easy to find with help from the catalogue sales staff.

The crown molding on the bookcase is the last complex work to complete. It is made from a solid piece of walnut with finished measurements of 2 inches × 4.5 inches × 6 feet. You selected the board from the available stock at the local specialty hardwood lumber supplier. It has promising features visible in the grain on the finished sides and on the end of the board, and you are looking forward to having the entire piece ready for finishing.

You followed Lonnie Byrd's recommendation and built the secretary from American black walnut, an exotic species. Typical prices for this variety are three to four times that

Case Studies in Mechanical Engineering: Decision Making, Thermodynamics, Fluid Mechanics and Heat Transfer,
First Edition. Stuart Sabol.
Companion website: www.wiley.com/go/sabol/mechanical

Figure 12.1   Philadelphia low boy.

for furniture-grade yellow pine. Your investment of time and money in the single piece has been considerable. In addition, completion of the molding requires special router bits specifically designed by Lonnie Byrd. After six months in the shop, your spouse has grown weary of seeing this project take all your time. Counting the many trips to the lumber yard, numerous orders of specialty hardware, special tools, finishing supplies and consumable products you have reached the endpoint on time and spending. There can be no mistakes on the crown molding.

Looking through the instructions, the crown molding requires five separate cuts and a considerable amount of handwork to smooth the machining marks. The second cut depends on correctly completing the first, the third requires a perfect second, and so forth. Each cut has a number of degrees of freedom, each of which could lead to serious problems and perhaps result in starting over. The task seems overwhelming but without the crown molding the work has an unfinished look. Commercially available moldings do not fit or would be unsightly. There are no options but to complete the designed molding with the material in hand.

After completing the first cut, a diagonal rip between opposite corners the length of the board, you are ready to set up the table saw to cut the central feature and focal point of the molding – a cove. A cove is a concave, circular feature in the wood that, in commercially produced applications, is made with a shaper. Such a tool is out of reach for most amateur woodworkers. However, a cove can be completed on a table saw by directing the piece diagonally across the blade. Small cuts are made, slowly deepening the cove to the desire depth. The angle of approach to the blade and depth of cut determine the width of the cove.

The *Fine Woodworking* article suggests sighting across the blade to set the board at the correct angle. You are not an expert marksman, and have never been good at shooting pool, so this option would be subject to error. An Internet search turned up a book containing setup angles for discreet coves. The book was published privately a number of years ago and must be purchased directly from the author. Besides the risk that the author has moved, you cannot be sure the right settings are included. Waiting for the book to be delivered could impact your motivation to complete the secretary.

The importance of the cut demands a more accurate method with fewer degrees of freedom. At first glance, it appears that a bit of engineering could solve the problem.

12.1   Background

Eighteenth-century furniture includes some of the best examples of design and craftsmanship. Considering the tools available at the time, each piece of furniture must have required master skills from many different crafts. Today, power tools make short work of once labor-intensive tasks but still require accurate and precise techniques to reproduce the look and structural integrity of eighteenth-century furniture. For this project – crown molding – three power tools are employed: the band saw, table saw, and router.

12.1.1   Band Saw

A band saw, such as the one pictured in Figure 12.2, uses a metal blade made into a continuous band with cutting teeth along one edge. Two or more wheels spinning in the same plane with the saw blade provide a continuous movement of the blade for straight or curved cuts through wood or metal. The work piece rests on a table through which the blade passes. The upper blade guide may be raised or lowered accommodating various work piece sizes, and the table can tilted to provide a variety of options for structural or decorative shapes. A fence can be mounted on the table to aid with straight cuts through the work piece.

Band saws may be mounted on a table or the floor. Hand-held varieties are often used in metal work for cutting through smaller sections of pipe or structural steel members.

There are several standard circumference sizes for blades. The longer the circumference the wider the opening through which the work piece may pass.

12.1.2   Table Saws

A table saw, Figure 12.3, is a workhorse in the wood shop. For large productions requiring many similar cuts, single cuts, shaping of tenons, miter cuts, patio, or fine furniture, the table saw is a major part of the project. The table is usually made of cast iron, ground smooth and flat. The table saw blade may be raised through the table and tilted up to 45°.

A cross cut miter gage slides through slots in the table parallel to the blade for cross cutting lumber. The miter may set at angles up to plus and minus 60° for cross cuts, miter cuts, or compound angle cuts working in combination with blade tilt. There are two conventions for miter gage angles, either 0° or 90° in the center. A miter gage may have preset stops for commonly used angles.

The standard fence is set parallel to the blade for rip cuts of long boards, and may be set with a spacer block for repetitive cross cutting. Cutting is always accomplished with either the crosscut miter or a fence, never both, and never by hand without a properly installed guide. Cross cutting is never done with the fence unless there is a spacer block that provides ample space between the board and fence.

Figure 12.2   Floor mounted band saw.

Figure 12.3   Floor model table saw.

Accurate initial setup of the saw is critical to ensure the fence and cross-cut guides are parallel to the blade, and the blade angle stops at 0° and 45° tilt are accurate. Regular application of bowling alley wax on the table prevents rusting and ensures smooth operation of the cross cut guide. Floor models collect dust and cuttings in the base, which can be evacuated to help keep the work area clean and reduce fugitive dust for a safer workshop.

In the United States, the standard table-saw blade is 10" in diameter, and the cutting teeth have a kerf of 0.125" – the width of the tooth. Blades may be purchased for specific type of cuts, but a typical all-purpose blade has teeth with alternating bevels.

12.1.3 The Router

A router (see Figure 12.4), in woodworking, is a tool to remove wood on the face or surface of a work piece. The router has a variable, high-speed spindle (up to 30 000 rpm) fitted with a cutting bit for straight or decorative designs. A router may be operated by hand using fixtures and guides to steady the power tool, or it can be mounted in a table. Table mounting permits accurate depth of cut, and a stable frame for smooth, precise forming of shapes for furniture, cabinetry, or structural shapes. A plunge router allows the router to be started prior to inserting into the work piece. Freehand use of the router is rare due to the high-speed spindle and the tendency of the router to wander through the work piece. The router bit may be fitted with a guide bearing such as the one shown in Figure 12.4.

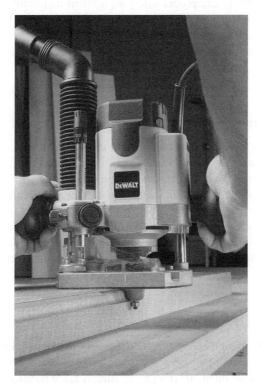

Figure 12.4 Plunge router with guide bearing on bit. *Source*: Reproduced by permission of Black & Decker US, Inc.

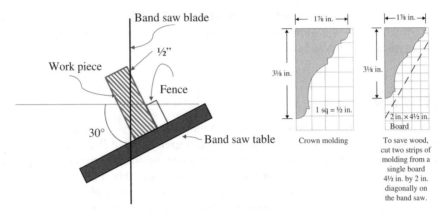

Figure 12.5 Band saw setup.

12.1.4 Safety

In all cases, power tools can be dangerous. Improper operation, careless actions, loose-fitting clothing, or failure to use proper safety equipment including eye and ear protection can lead to serious accidents and permanent losses of hearing, sight, or appendages. Prior to use, read and study the safety precautions, mount the safety precautions in a prominent location in the shop, seek expert advice prior to use, and receive training from a qualified instructor.

12.1.5 Measurements

Lumber is sold in sizes based on the rough cuts from the tree prior to drying and planing to finished dimensions. Finishing generally removes ¼" for sizes under 2", ½" for sizes between 2" and 8" and ¾" for sizes above 10". Hardwood lumber is sold by the board-foot (1" × 12" × 12", or 144 in³) in the United States and Canada. In the European Union, hardwood lumber is generally priced by the length in meters with unit costs dependent on thickness and width. European sizes are in millimeters and are nominally similar to US sizes.

Rough-cut lumber is generally available at specialty lumberyards. The cost is less than finished lumber, but it requires planning prior to use in furniture and most structural designs.

12.2 Case Study Details

The crown molding shape for the Pennsylvania Secretary is shown in Figure 12.5. It is formed by first making a diagonal cut through a 2" wide x 4.5" tall board. The first cut is made with the band saw with the table tilted 30° from horizontal. A fence parallel to the blade is set such that the blade cuts the board along a diagonal cutting the 4.5" sides ½" from each corner – see Figure 12.6. A wide blade, 0.375" or greater, helps keep the blade vertical through the cut and

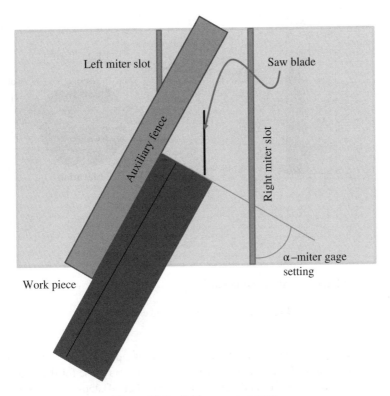

Figure 12.6   Table saw cove setup.

maintain a straight cut through the length of the board. A wide blade also permits greater tension on the blade improving the quality of the cut. Ensure you use a new, sharp blade. A dull blade tends to arc through the cut.

12.2.1   Exercise

1. Using a compass and a straight edge, find the center of the circle forming the cove in the crown molding shown in Figure 12.5. Hint: perpendicular bisectors of two chords of the circle will meet in the center of the circle.
2. Scale the depth of the cut and the width of the cove.

12.2.2   The Cove

As described earlier, the cove is cut on the table saw by aligning the work piece at an angle to the blade so that an arc is cut into the piece – see Figure 12.6. The *Fine Woodworking* article suggests drawing the cove on the end of the work piece and then sighting across the blade, making adjustments in the angle, until you see that the blade will cut the desired arc.

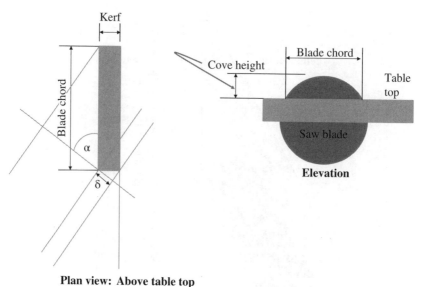

Plan view: Above table top

Figure 12.7   Setup, plan and elevation.

This sounds rather approximate. An Internet search turns up a variety of techniques, most using some type of fixture. There are even products you can purchase for the setup.

The processes shown in these methods require a number of steps – building the fixture, setting the fixture, measuring an angle, then setting a fence at the measured angle at the required distance from the blade for the actual cut. Each step adds another potential source of error and uncertainty in the final cut. Additionally, the last thing you need is another fixture in your small shop.

Looking again at Figure 12.6, the answer is apparent. The projection of the blade onto the work piece is simply one leg of a right-angle triangle with the blade as the hypotenuse – see Figure 12.7.

The one complexity is that the blade has a thickness – the kerf. The back side of the blade adds a minor distance to the width of the cove, due to the three-dimensional shape of the blade – see Figure 12.7. But this is just a bit more trigonometry.

Exercise

1. Write the equation for the chord (c) of the table saw blade struck against the table as a function of the cove height – h.
2. Write the equation for δ in Figure 12.7 as a function of the auxiliary fence angle.
3. Write the equation for the cove width as a function of the blade chord, δ, and of the auxiliary fence angle (α) shown in Figure 12.7.
4. Using the results of Exercise 2. above determine the proper auxiliary fence angle for the cove of the crown molding. Test your answer at: www.finewoodworking.com/how-to/article/cove-angle-calculator.aspx (accessed February 2, 2016).

Figure 12.8   Asymmetric cove on the table saw.

12.2.3   Extra Credit

Asymmetric coves, Figure 12.8, are a class of coves with the apex or crest of the cove displaced from its center line. Asymmetric coves can be cut on the table saw by tilting the saw blade. Write the general equations necessary to cut any type of cove, symmetric or asymmetric on a table saw.

12.3   Closure

The crown molding was completed and installed as conceived. It was the perfect design element to finish the bookcase providing the right balance of detail and size (Figure 12.9). Settings calculated by a simple spreadsheet resulted in a finished cove just a few thousandths of an inch from expected height, width, and location within the work piece. Additionally, no new fixture was required.

The author submitted a "method of work" to *Fine Woodworking,* which was written as a feature article with Mark Schofield, appearing in the February, 2004, edition, No. 168. It is available as a reprint and was included in two volumes of related articles published by Taunton Press, which can be ordered from FineWoodworking.com. The spreadsheet was rewritten as a web application with an iterative solution for an auxiliary fence angle and blade tilt to yield the desired cove dimensions for symmetric and asymmetric coves. The web application is a permanent part of the *Fine Woodworking* web site.

Hobbies are good. To have the opportunity to contribute to the science of a pastime that I enjoy was an honor. Increasing the level of engineering in the art of woodworking, or any profession, should become a regular endeavor for engineers who enjoy their work and hobbies.

Modern computational capabilities and international collaboration have the potential to create new work methods that are more accurate, easier to apply, safer, and more affordable

Figure 12.9   Pennsylvania secretary.

than anything in the present-day market. If you ever have the opportunity to contribute to your chosen field of interest or to your hobby, do not hesitate.

The secretary's eighteenth-century design continues to be an ideal, modern workstation for a laptop in addition to being a handsome addition to our home (Figure 12.10). It was a lot of fun to build, and great fun to write about.

Figure 12.10   Secretary gallery.

12.4   Glossary

Bit: a metal cutting tool mounted in a rotating spindle of a power tool.
Cross cut: a cut through lumber made at a right angle to the wood grain.
D: table saw diameter.
h: cove height.
k: kerf.
kerf: width of a cut.
miter: center point measurement for the cross cut miter gage, either 0° or 90°.
Rip cut: A cut through lumber in the direction of the wood grain.
w: width of the cove.
α: auxiliary fence angle measured on the cross-cut miter.
δ: length of the blade chord due to the saw blade kerf.
θ: blade tilt angle (0°–45°).

12.5   Solutions

Section 12.2.1

1. See Figure 12.11.
2. h = 5/8 inches (15 mm);   w = 2 3/8 inches (60 mm).

Section 12.2.2

1. $c = 2\sqrt{Dh - h^2}$
2. $\delta = k \cdot \{ IF\ miter = 0\ then\ \sin(\alpha)\ else\ \sin(90 - \alpha) \}$
3. $w = c \cdot \cos(\alpha) + \delta$
4. c = 4.893″, 124.3 mm; δ = 0.111″, 2.821 mm; α = 63°

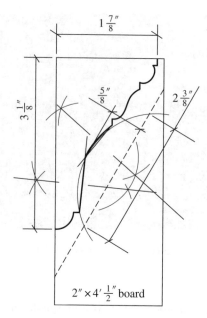

Figure 12.11   Scaling cove dimensions.

Section 12.2.3: Extra Credit

Tilting the blade from vertical requires more of the blade to project from the table to meet the required cove height. The tilt is projected onto the plan of the cove similar to the chord length; therefore, the apex of the cove moves off center creating an asymmetric shape. Additionally, the kerf is further exposed; thus, slightly increasing the blade chord. Some saws tilt to the left, others to the right. The result is the same for both types, but requires opposite-hand setup for the saw to prevent feeding the work piece over the top of a tilted blade. For a center miter reading of 0°, the results of 12.2.2 above become the expanded version below for a blade tilt angle of θ:

1. $c = 2\left[\dfrac{Dh}{\cos(\theta)} - \left(\dfrac{h}{\cos(\theta)}\right)^2\right]^{0.5}$

2. $\delta = \dfrac{k}{\cos(\theta)}\{IF\ miter = 0\ then\ \sin(\alpha)\ else\ \sin(90-\alpha)\}$

3. $w = c \cdot \cos(\alpha) + \delta$

4. $tilt\ from\ vertical = h \cdot \tan(\theta)$

5. $Apex\ of\ fset = tilt\ from\ vertical \cdot IF\{miter = 0,\ then\ \cos(90-\alpha)\ else\ \cos(\alpha)\}$

Further Reading

Woodworking magazines are a good source of information on plans, techniques, reviews of power tools, safety procedures, new and innovative solutions, and so forth. The ToolCrib blog (http://www.toolcrib.com/blog/2008/07/the-top-5-woodworking-magazines-and-the-22-runners-up, accessed February 2, 2016) ranked the leading magazines from readers of the blog, with the following results for the top five:

1. *Fine Woodworking*
2. *Wood Magazine*
3. *Popular Woodworking*
4. *Shop Notes*
5. *Wood Smith*

The magazines typically target a specific skill level, so it is a good idea to review a copy from a news stand prior to subscribing. A few books the author found useful include the following:

References

Charron, A. (1997) Shop-Tested Outdoor Furniture You Can Make, Meredith Corp., Des Moines, IA.
Charron, A. (2000) Desks, Taunton Press, Inc., Newtown, CN.
Greene, J. P. (1996) American Furniture of the Eighteenth Century, Taunton Press, Inc., Newtown, CN.
Yoder, R. A. (1985) Making Period Furniture, Taunton Press, Inc., Newtown, CN.
Yoder, R. A. (ed.) (1999) Handcrafted Cabinetry, Reader's Digest Association, Inc., New York, NY.
Yoder, R. A. (2005b) Working with Tablesaws, Taunton Press, Inc., Newtown, CN.
Yoder, R. A. (2005a) Powertool Techniques, Taunton Press, Inc., Newtown, CN.

Appendix

Steam-table functions are available in Excel at no cost from http://www.mycheme.com/steam-tables-in-excel/(accessed February 2, 2016) and www.me.ua.edu/me215/f07.woodbury/ExcelStuff/XSteam-v2a.xlsm (both accessed February 2, 2016). These functions, for the 1997 International Association for the Properties of Water and Steam (IAPWS), were developed by Magnus Holmgren. The functions are provided "as is," and no responsibility is taken for any errors in the code or any damage resulting from their use. Individuals are free to use, modify, and distribute the functions as long as authorship is properly acknowledged. Commercial applications require notification, as shown in the Visual Basic module.

To use the steam table functions, open the Excel file provided from the download. In Excel 2013, go to the "Developer" tab on the ribbon menu at the top of the screen. If the Developer menu is not available, click on "File" then "Options" and "Customize Ribbon." Add the Developer menu as shown in Figure A.1.

On the Developer tab click on "Visual Basic" to reveal the coding for the functions. Along the left-hand side of the screen, open the "Modules" to show the list of modules included with the steam-table functions. See Figure A.2. As shown, there are two modules: "Extra_Functions" and "X_Steam_Tables."

Highlight each function, one at a time, by clicking once. Then click on "File" and "Export." Choose a file location where you will be able to save and retrieve the modules for future use.

To bring the functions into a file, click on "File," "Import" from the "Visual Basic" menu item on the "Developer" tab of Excel.

To return to the Excel file, simply click on the Excel icon on the menu bar.

The available functions are shown on the function tab of the spreadsheet with the required units. When the modules have been imported into the file, these functions are available and

Case Studies in Mechanical Engineering: Decision Making, Thermodynamics, Fluid Mechanics and Heat Transfer,
First Edition. Stuart Sabol.
© 2016 John Wiley & Sons, Ltd. Published 2016 by John Wiley & Sons, Ltd.
Companion website: www.wiley.com/go/sabol/mechanical

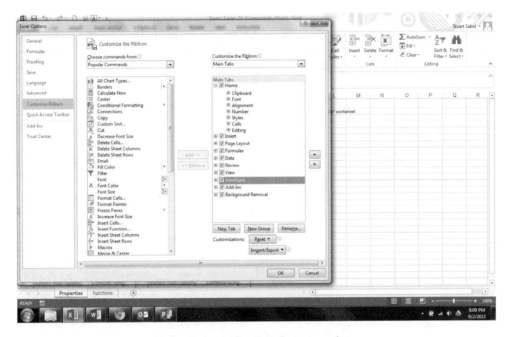

Figure A.1   Adding Developer menu item.

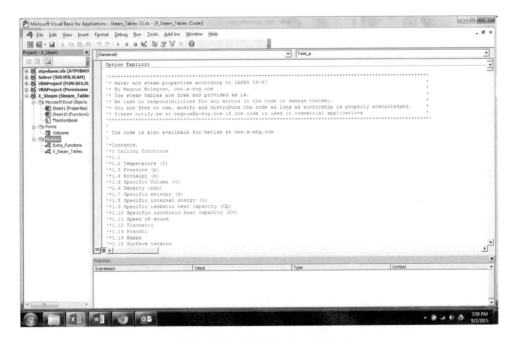

Figure A.2   Visual Basic modules.

can be used as any standard function provided with Excel. Function calls may be typed directly into a cell, or the "Formulas" menu on the ribbon can be used. To show the steam table functions, select the "All" option in the function category on the "Insert Function" dialog box, and find the function in the alphabetical listing.

Glossary

ActiveX: software framework created by Microsoft. ActiveX controls can be inserted into Excel spreadsheets using the "Insert" menu on the "Developer" tab. See Figure G.1.

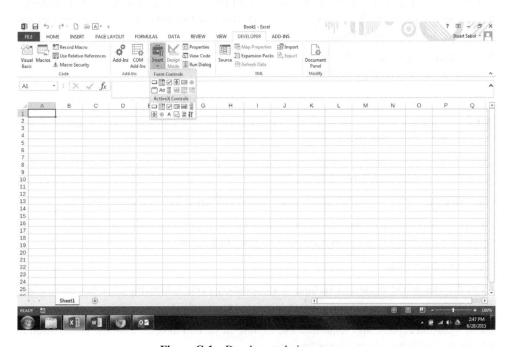

Figure G.1 Developer tab, insert menu.

Case Studies in Mechanical Engineering: Decision Making, Thermodynamics, Fluid Mechanics and Heat Transfer,
First Edition. Stuart Sabol.
© 2016 John Wiley & Sons, Ltd. Published 2016 by John Wiley & Sons, Ltd.
Companion website: www.wiley.com/go/sabol/mechanical

Aero-derivative: a modified gas turbine used for aircraft with a power turbine used to drive an electric generator.

AF: availability factor. The hours available to operate at maximum capacity divided by the period hours – equation (G.1):

$$availability\ factor = \frac{operating\ hours - equivalent\ outage\ hours}{period\ hours} \tag{G.1}$$

Agent: typically a chemical substance used by a military to perform a specific task.

API: American Petroleum Institute: 1220 L Street NW, Washington, DC 20005-4070.

ASME: American Society of Mechanical Engineers, headquartered at Two Park Ave. New York, NY 10016-5990.

Blowdown: when water is evaporated, boiled, dissolved solids in the water are concentrated in the remaining liquid. To maintain water quality at acceptable levels, a portion of the water must be discarded to carry away a quantity of dissolved solids equal to the quantity of dissolved solids entering the boiler system. This discharge is termed "blowdown."

BOD: basis of design. An engineering document that is progressively developed through a project design to establish and document the system requirements, design assumptions, and foundation for the system design.

Btu: British thermal unit. The energy required to cool or heat one pound of water by 1 °F. Internationally accepted as 1055.06 J.

Bucket: a manufacturer's term for a rotating blade in axial-flow turbomachinery.

BWR: boiling water reactor. A nuclear power cycle in which steam for power generation is manufactured inside the nuclear reactor.

Cal: calorie. The energy required to raise one gram of water one degree centigrade. The mean value is approximately 4.19002 J.

Category M: Piping design category defined by ASME B 31.3 for lethal service.

CBM: condition-based maintenance. A maintenance protocol utilizing measurements of equipment condition to perform maintenance prior to a failure.

CC: combine cycle.

CCGT: combined cycle gas turbine – a combined cycle consisting of a gas turbine and steam turbine.

CCS: carbon capture and sequestration, capture of carbon dioxide from the combustion of a fossil fuel with permanent sequestration.

CED: cause-and-effect diagram. A modified P&ID showing the instrumentation and control logic for a system.

CF: capacity factor. The average or instantaneous output expressed as a percentage of the maximum dependable capability or design capacity of a system or process. For a power-generating facility, capacity factor may be calculated as the ratio of the actual generation (MWh) in a period to the product of the period hours and the maximum dependable capability (MW).

CHP: combined heat and power. A power plant providing a portion of its steam generation for heat. Also referred to as cogeneration.

CI: combustion inspection. A maintenance inspection of the gas turbine combustion section.

Cogeneration: see CHP.

Combined cycle: tandem thermodynamic cycles using a common heat source.

Cooling tower: generally a direct contact heat exchanger that cools warm water from waste heat rejection against air by evaporating a portion of the warm water. Cooled water exits the base of the tower at a temperature slightly above the dew point of the incoming air. Air leaves the top of the tower through exhaust fans at 100% relative humidity and carries a small amount of liquid droplets referred to as "drift." Makeup water is added to the cooling tower basin to replenish evaporation and blowdown. Due to evaporation, dissolved solids in the water stream become concentrated and must be removed to prevent accumulation of solids. A portion of the warm water from the waste heat rejection, "blowdown," is discarded to perform this function.

Cross compound steam turbine: a steam turbine with two or more casings, each with a series of blades and vanes driven by the same source of steam, the axils of which drive separate generators.

CWC: Chemical Weapons Convention. An international treaty drafted in 1992 banning the use, manufacture and storage of chemical weapons.

Desuperheater: a heat exchanger designed to transfer heat from a superheated vapor to a temperature slightly above its saturation temperature.

DOD: domestic object damage. Internal damage caused by an object originating within a gas turbine.

EIA: US Energy Information Administration.

EOH: equivalent outage hours. The full outage hours equivalent to the actual hours of a forced partial reduction in output:

$$EFOH = \frac{forced\ partial\ outage\ hours * size\ of\ reduction}{maximum\ dependable\ capability} \tag{G.2}$$

EPC: engineer, procure, construct. A contracting strategy whereby a single entity provides the detailed engineering, equipment and material procurement, and construction at a single price. Most often the EPC contract has a fixed price but may include incentives, penalties, and escalation clauses depending on the complexity and duration of the project. Owners employ EPC contracting strategies to transfer risks associated with schedule, equipment, and material prices, supply and logistics, labor, and so forth, from the owner to the EPC contractor. On large, complex projects, the EPC contractor may be a consortium of vendors that have agreed to work with the owner under the terms of the contract.

EPRI: Electric Power Research Institute: 1325 G Street NW suite 1080, Washington, DC 20005.

Equivalent fired hours: total fired operating hours plus equivalent hours, defined by the manufacturer, which account for additional stress on hot gas path components. Equivalent hours may include startup and shutdown, trips, use of liquid fuel, use of steam for power augmentation, and so forth.

ERCOT: Electric Reliability Council of Texas.

Excess air: a practical quantity of air provided in excess of the stoichiometric requirement to combust the fuel to near extinction. Usually expressed as a percentage of the stoichiometric requirement. The excess air quantity can be optimized for the best overall efficiency by balancing losses due to incomplete combustion with the losses due to dry gas leaving the stack at a temperature above the ambient.

FC: fixed costs. A category of costs that are independent of the production rate. Generally includes the labor, fixed cost service contracts, insurance, rents, and so forth.

FEED: front-end engineering design, broadly defined as the engineering design required for a firm price quotation supplied for detailed design, engineering, procurement and construction.

FEL: front-end loading. Conceptual engineering, permitting, planning, and project economic analysis completed prior to FEED. Front-end loading may be divided into several phases with each phase generating greater detail of the design and planning requirements.

Five-X: decontamination level permitting release of items exposed to chemical warfare munitions from the army's control.

FOD: foreign object damage: internal damaged caused by an object originating outside a gas turbine.

FOR: forced outage rate. The percent of time a facility or piece of equipment is forced out of service – equation (G.3):

$$FOR = \frac{forced\ outage\ hours + EFOH}{period\ hours} \tag{G.3}$$

Fortran: A computer language developed by IBM in the 1950s for scientific applications.

FSAR: final safety analysis report, a required application document for the NRC's permit approval for a nuclear power plant.

GHG: greenhouse gases contributing to man-made climate change, including carbon dioxide, methane, nitrous oxide, and others.

HD: distilled mustard ($C_4H_8Cl_2S$). A blistering chemical warfare agent.

Heat rate: the ratio of the fuel energy to the generated electrical output of a power-generating facility. The ratio is expressed most often as kJ/kWh, GJ/MWh, Btu/kWh or MMBtu/MWh depending on the location and audience.

HEI: Heat Exchange Institute: 1300 Sumner Ave. Cleveland, OH 44115.

HEPA: high efficiency particulate air: a type of high efficiency air filter first employed during the Manhattan Project in the 1940s.

HGPI: hot gas path inspection. A maintenance inspection of the gas turbine hot gas path.

HHP: high-high Pressure. Terminology generally reserved for supercritical steam power cycles. The HHP steam system would be located upstream of the first reheater in a multiple reheat cycle.

HHV: higher heating value: the fuel heat of combustion measured by bringing the products of combustions to the precombustion temperature and condensing any water vapor produced during combustion. Fuels are normally purchased, and priced on an HHV basis.

Hot gas path: gas turbine sections exposed to combustion gases, including the combustors, transitions and turbine.

HP: high pressure. The highest pressure of a subcritical steam-power cycle.

HRSG: heat recovery steam generator. A heat exchanger utilizing waste heat, usually from a gas turbine or diesel engine, to convert water into steam or superheated steam for another use. Steam generation is commonly used to drive a steam turbine and provide heat for chemical processing, crude oil refining, food preparation, building heat, or other uses. Heat-recovery steam generators may be physically configured with horizontal gas flow with heat-transfer tubes installed vertically or with vertical gas flow and horizontal tubes. Most HRSGs are used in subcritical steam cycles with one or more steam drums of various pressure levels to separate saturated liquid from vapor with natural circulation evaporators. "Once-through steam generators" (OTSG), are occasionally used. Once-through designs do not have a steam drum and may produce a steam / water mixture, which is separated externally from the OTSG.

IAPWS: International Association for the Properties of Water and Steam.

IGCC: integrated gasification combined cycle; solid fuel gasification, usually a partial combustion of coal to form carbon monoxide as a fuel for a CCGT.

IGV: inlet guide vanes. The initial stationary vanes at the inlet of a gas turbine. These vanes have adjustable settings to control the air flow and, therefore, the output of a gas turbine.

Inlet fogging: A gas turbine inlet air treatment for cooling the inlet air to near the dew point by evaporating a water spray.

IP: Intermediate pressure. Usually refers to the steam turbine casing or pressure level immediately downstream of the reheater.

IPP: independent power producer. A merchant power producer operating independently of a regulated or state-owned utility.

LCOE: levelized cost of electricity. The ratio of the present value of the lifetime cost of building and operating the facility to the present value lifetime energy production.

Level A: US Occupational Safety and Health Administration level of personnel protection required for the most severe hazards with skin, eye, and respiratory risks.

Level A area: containment area known or suspected to have hazardous compounds that may be lethal.

LHV: lower heating value: the fuel heat of combustion neglecting the heat of vaporization of water produced by combustion. Normal industry practice is to base the efficiency of a gas turbine on the LHV fuel value.

LMTD: log mean temperature difference. For a general counter flow heat exchanger this quantity is defined by equation G.4 with definitions shown in Figure G.2:

$$LMTD = \frac{\left(T_{H1} - T_{C2}\right) - \left(T_{H2} - T_{C1}\right)}{ln\left(\frac{\left(T_{H1} - T_{C2}\right)}{\left(T_{H2} - T_{C1}\right)}\right)} \tag{G.4}$$

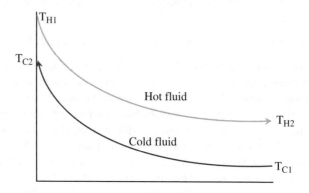

Figure G.2   LMTD definitions: counter flow.

where:

T: temperature;
C: cold fluid;
H: hot fluid;
1: inlet;
2: exit.

LNG: liquefied natural gas. Natural gas cooled to a liquid state, generally near ambient pressure.

Long-run marginal cost: The cost for a new entry into a market. The long-run marginal cost includes the owner's development cost, cost of construction, and the fixed and variable operating costs over the life of the facility.

LP: low pressure. Usually refers to the steam-turbine casing or pressure level at the intermediate pressure steam-turbine casing exhaust.

LPO: lost profit opportunity. An LPO is generally expressed as a monetary value equal to the gross or net margin that could have been produced during a forced or partial forced outage. How an LPO is expressed may depend on the corporation.

MDC: maximum dependable capability.

MI: major inspection: overhaul outage for the gas-turbine compressor, combustion, and turbine sections.

MM: Roman numerals for 1 000 000.

Monte Carlo simulation: a method of simulating complex systems with many degrees of freedom through the use of random variables. First invented by Stanislaw Ulams at the Los Alamos National Laboratory.

NEHR: net effective heat rate. A value of heat rate calculated to account for heat energy supplied by a CHP plant. There are two common forms of NEHR. One calculation treats the compensation paid for heat energy as a reduction of the quantity of fuel required for power production. The other common form of the equation treats the heat energy as a second output in the same units as electric output (kWh or MWh.) Either treatment is acceptable and generally defined for a specific project in an agreement for the purchase and sale of electricity and steam.

NERC: North American Electric Reliability Council.

NO_x: oxides of nitrogen, NO and NO_2.

NPSH: net positive suction head. Pressure entering a pump measured in length of the fluid pump above the saturation pressure of the liquid.

NPV: net present value.

$$NPV = \sum_{t=1}^{n} \frac{\left(R_t - C_t\right)}{\left(1+r\right)^t} \tag{G.5}$$

where:

C_t = cost at time t;
n = periods beginning at 1 through the project life;
r = discount rate (fraction/period);
R_t = revenue at time t;
t = periods 1 through n.

NRC: United States Nuclear Regulatory Commission, an independent government agency created in 1974. The National Research Council, a division of the US National Academy of Science.

O&M: operations and maintenance.

PAG: steam power augmentation. Steam injected beyond the combustion zone of a gas turbine to increase power output.

P&ID: piping and instrument diagram: an engineering drawing of the piping systems of a process including the piping line sizes, pipe schedules, piping materials, instrumentation, equipment, insulation, heat tracing, equipment elevation requirements, piping slope requirements, valves, valve types, pressure relief value set points, and others. Detailed P&IDs include all line, valve and equipment nozzle sizes, and vendor designations for equipment connections. P&IDs are prepared prior to the design of the three-dimensional piping design and isometric drawings. Detailed piping isometrics may result in minor changes to P&IDs prior to their issue for construction.

PC: pulverized coal. Coal ground to a fineness such that 80% will pass through a 200 mesh screen, 200 wires placed horizontally and vertically per inch.

PDP: process design package. An engineering design package containing the chemical engineering design, basic process control logic, and process equipment requirements typically produced in FEL.

Peaker: an electric generating plant used during periods of high electric prices or sudden electric power shortages. Peaker plants, also called **peaking** plants, are typically gas turbine or diesel generators. However, gas-fired Rankine cycle steam power plants may be converted for peaking service.

PFD: process flow diagram. A diagrammatic drawing of the chemical process, usually produced with corresponding tables of fluid properties, flows, pressures, and temperatures of the process streams shown on the diagram.

pH: Negative $\log_{10}$ of the hydrogen ion concentration in solution. A numeric measure of the acidity or alkalinity of a solution.

Planned outage rate: The percentage of time a facility or piece of equipment is scheduled to be outage of service for maintenance, also referred to as the planned outage factor. Refer to Appendix F of the NERC reporting guidelines available at: http://www.nerc.com/pa/RAPA/gads/Pages/Data%20Reporting%20Instructions.aspx (accessed February 2, 2016).

PMBOK®:   Project Management Book of Knowledge developed by PMI.

PMI: Project Management Institute: 14 Campus Blvd. Newtown Square, PA 19073-3299.

Preventive maintenance: A time-based or run-time-based maintenance protocol in which maintenance occurs on a regular schedule.

ppmvd: parts per million on a volume basis in a dry (moisture-free) gas.

PTC: Performance Test Code, published by ASME.

PUC: public utility commission, usually a state government agency charged with regulating public utilities, including electric utilities, within their jurisdiction.

PWR: pressurized water reactor. A nuclear power cycle with an intermediate pressurized water loop that exchanges heat between the nuclear reactor and the steam power cycle.

Quality assurance: practices followed to ensure the procedures put in place to assure quality products are being followed.

Quality control: measurements taken to determine if products meet specific standards of quality.

RCM: reliability centered maintenance. A programmatic engineering maintenance protocol intended to minimize maintenance while preserving system function.

Reactionary maintenance: maintenance occurring only after a failure.

Reboiler: a natural circulation heat exchanger that partially evaporates liquid from the bottom of a distillation column providing the heat energy that drives distillation.

Reheater: heat exchanger surface used to superheat steam from the exhaust of a steam turbine casing.

Reliability factor: the percent of time a facility or piece of equipment operates at full load during a period that does not include planned maintenance outages – equation (G.6).

$$reliability\ factor = \frac{operating\ hours - EFOH}{period\ hours - planned\ maintenance\ hours} \qquad (G.6)$$

REP: retail electric provider. A licensed reseller of electric power to retail customers within ERCOT.

RPS: renewable portfolio standard. A governmental mandate, or target, to achieve installation of a specific quantity, or percentage, of renewable generation for a portfolio providing electric service. The service area may be the territory served by a regulated utility, a state or an entire country.

SC: simple or single cycle.

SCR: selective catalytic reduction.

SGU: steam-generating unit. The secondary side of a pressurize water reactor generating steam for the nuclear steam cycle and providing cooling to the nuclear reactor. Water under pressure circulates between the nuclear reactor where it is heated and the SGU where it is cooled against steam used to power a steam-turbine generator.

Short-run marginal cost: the variable costs required to produce the next increment of a product.

SO_X: oxides of sulfur, SO_2 and SO_3.

Spark spread: the difference between the wholesale electric price and the short-run marginal cost to produce the electricity at a given heat rate. The heat rate may be a consensus value generally accepted to represent the market, the state-of-the-art technology to produce electric power, or the generator's nominal heat rate depending on the application.

Steam calorimetry: a calculation of the thermal input to the steam power cycle in a nuclear power plant.

Steam power augmentation: steam injected into the gas turbine immediately upstream of the turbine section to increase output.

Superheat: the temperature difference between the actual temperature of a vapor and its saturation temperature at the same pressure.

Superheater: a heat exchanger designed to increase the temperature of a saturated vapor.

Tandem compound steam turbine: a steam turbine with two or more casings, each with a series of blades and vanes driving an axil, the axils of which are joined end to end to drive a single generator, pump, or compressor.

TBC: thermal barrier coating. A multilayer coating on gas-turbine blades and vanes in the hot gas path used to insulate the base material. The top TBC layer is generally a ceramic oxide such as zirconium oxide.

Three-X: decontamination level permitting transport of items exposed to chemical warfare munitions.

ton: 2000 lb_m, also a **short ton**.

tonne: 1000 kg, also a **long ton**.

Tornado diagram: a bar chart with categories listed vertically and arranged such that the longest bars appear at the top of the chart. In Excel, a tornado diagram is construction from a "stacked bar chart" with two series spanning the listed categories.

Vane: a stationary airfoil in axial flow turbomachinery.

U: overall heat transfer coefficient. The coefficient accounting for resistance to heat transfer due to convection, conduction and fouling.

Variable costs: operating and maintenance costs that are proportional to, or a direct function of, the output level of a manufacturing facility.

Visual basic: third-generation programming language.

VOC: volatile organic compound.

Void fraction (ε):   dimensionless parameter representing the fraction of a two-phase fluid flowing as a gas or vapor.

Waterfall chart: a stacked bar chart with hidden columns used to show a progression to a total sum consisting of a series of inputs.

WBS: work breakdown structure. A hierarchal diagram organizing a project's scope – work products – into manageable segments.

Wet compression: a combination of inlet fogging and compressor water injection carried out in the inlet duct. A quantity of water in excess of that required to reach 100% relative humidity sprays into the inlet air stream resulting in liquid droplets of water entering the first stages of the gas turbine. These droplets evaporate in later compression stages, providing cooling and reducing power required for compression.

Index

Case Studies in Mechanical Engineering: Decision Making, Thermodynamics, Fluid Mechanics and Heat Transfer,
First Edition. Stuart Sabol.
© 2016 John Wiley & Sons, Ltd. Published 2016 by John Wiley & Sons, Ltd.
Companion website: www.wiley.com/go/sabol/mechanical